The Empirical Evidence on the
Efficiency of Forward and Futures
Foreign Exchange Markets

FUNDAMENTALS OF PURE AND APPLIED ECONOMICS

Fundamentals of Pure and Applied Economics is an international series of titles divided by discipline into sections. A list of sections and their editors and of published titles may by found at the back of this volume.

The Empirical Evidence on the Efficiency of Forward and Futures Foreign Exchange Markets

Robert J. Hodrick
Northwestern University, USA

A volume in the Balance of Payments and
International Finance section
edited by
William Branson
Princeton University, USA

harwood academic publishers
Australia • China • France • Germany • India • Japan •
Luxembourg • Malaysia • The Netherlands • Russia • Singapore •
Switzerland • Thailand •United Kingdom • United States

First published 1987
Third printing 1996

Emmaplein 5
1075 AW Amsterdam
The Netherlands

Library of Congress Cataloging-in-Publication Data

Hodrick, Robert J.
  The empirical evidence on the efficiency of forward
and futures foreign exchange markets.

  (Fundamentals of pure and applied economics,
vol. 24. Balance of payments and international
finance section; ISSN 0191-1708)
  Bibliography: p.
  Includes index.
  1. Forward exchange. 2. Foreign exchange. I. Title.
II. Series: Fundamental of pure and applied economics;
v. 24. III. Series: Fundamentals of pure and applied
economics. Balance of payments and international
finance section.
HG3853.1143 1987   332.4'3   87-17790
ISBN 3-7186-0415-9

Contents

Contents

Introduction to the Series

Drawing on a personal network, an economist can still relatively easily stay well informed in the narrow field in which he works, but to keep up with the development of economics as a whole is a much more formidable challenge. Economists are confronted with difficulties associated with the rapid development of their discipline. There is a risk of "balkanization" in economics, which may not be favorable to its development.

Fundamentals of Pure and Applied Economics has been created to meet this problem. The discipline of economics has been subdivided into sections (listed at the back of this volume). These sections include short books, each surveying the state of the art in a given area.

Each book starts with the basic elements and goes as far as the most advanced results. Each should be useful to professors needing material for lectures, to graduate students looking for a global view of a particular subject, to professional economists wishing to keep up with the development of their science, and to researchers seeking convenient information on questions that incidentally appear in their work.

Each book is thus a presentation of the state of the art in a particular field rather than a step-by-step analysis of the development of the literature. Each is a high-level presentation but accessible to anyone with a solid background in economics, whether engaged in business, government, international organizations, teaching, or research in related fields.

Three aspects of *Fundamentals of Pure and Applied Economics* should be emphasized:

—First, the project covers the whole field of economics, not only theoretical or mathematical economics.

—Second, the project is open-ended and the number of books is not predetermined. If new interesting areas appear, they will generate additional books.

—Last, all the books making up each section will later be grouped to constitute one or several volumes of an Encyclopedia of Economics.

The editors of the sections are outstanding economists who have selected as authors for the series some of the finest specialists in the world.

J. Lesourne   *H. Sonnenschein*

The Empirical Evidence on the Efficiency of Forward and Futures Foreign Exchange Markets

ROBERT J. HODRICK

Kellogg Graduate School of Management, Northwestern University and National Bureau of Economic Research

1. INTRODUCTION

This monograph is a critical review of the empirical literature that studies the efficiency of the forward and futures markets for foreign exchange. The importance of this topic to government policymakers, firms, investors, and economists and the ready availability of data have spawned a large volume of research in this area. Although much work has been done, there does not appear to be a strong consensus either in the opinions of those who have conducted the research or in those who have read it. That this difference of opinion can be readily traced to strong prior opinions and relatively uninformative data is not surprising. Yet, we have learned a lot, and my purpose here is to lay out, in a systematic way, what I think we have learned.

The importance of this topic for policymakers stems from their need to assess the performance of alternative international financial systems. The existence and efficiency of organized markets for future delivery of foreign currencies was a critical link in early arguments for flexible exchange rates. Friedman [78] argued that the existence of such markets provided international traders a convenient means of hedging risks of currency fluctuations. In contrast, Nurske [167] argued that flexible exchange rates would be inherently volatile and subject to the whims of speculators.

The volatility of exchange rates, since the abandonment of the Bretton Woods arrangement in the early 1970's, has proven surprising to many economists. Indeed, the results of Meese and

1

Rogoff [158, 159] indicate that current economic models of spot exchange rate determination are generally unable to explain the movement in major currency exchange rates. Yet, without economic models of the determination of exchange rates, how can we know whether we need fixed rates or flexible rates or more or less intervention in foreign exchange markets? How can we ascertain what is a fundamental disequilibrium that would trigger a change in a fixed-rate system? The studies described here offer a scientific foundation upon which we can hope to develop answers to these questions.

Corporations and investors obviously care about the efficiency of forward and futures markets for foreign exchange. Whenever uncovered foreign currency investments are made, one component of the domestic currency return is any change that occurs in the domestic value of the foreign currency. The expected rate of change of the exchange rate contributes to the expected return and an assessment of this must be made to determine expected returns on foreign investment. A substantial amount of resources flow each year into forecasting exchange rates, and the evidence presented here has a direct bearing on such activities. Furthermore, contentions of unexploited profit opportunities and insufficient or overabundant speculation are rampant in this literature, and the evidence on various trading strategies and filter rules is presented below.

Finally, this survey is of interest to financial economists because it collects together a number of recent approaches to testing asset pricing models that may be usefully employed in other contexts. In this regard, a major thrust of the analysis below is that asset pricing specification tests ought to allow for the presence of conditional heteroscedasticity. Estimation procedures that allow for conditional heteroscedasticity and ones that model it are discussed.

The monograph is divided into seven remaining sections. Section 2 draws on the Lucas [145] model and the arguments of Cox, Ingersoll and Ross [34] to examine a theoretical model of the determination of forward and futures prices for foreign exchange. Since the Lucas model is a complete dynamic two-country general equilibrium model, it provides a useful starting point for the discussion. Unfortunately, since the purpose of the monograph is to survey the empirical literature, alternative theoretical frameworks cannot be addressed in the available space. Readers who desire more theoretical material ought to refer to the survey by Adler and Dumas [2] and the contribution of Branson and Henderson [24] in the excellent volume edited by Jones and Kenen [118]. Both of these articles do a good job of describing the state of the art in the theoretical modeling of international asset pricing.[1]

The empirical literature in this area has also been the subject of some recent surveys that ought to be read in conjunction with this monograph. Particular mention must be made of the contribution of Levich [133]. Levich surveys models of spot exchange rate determination as well as forward rates. The survey of Boothe and Longworth [21] also provides a different perspective and interpretation of the empirical literature on the efficiency of forward markets.

The major theoretical implications derived in Section 2 are that neither forward rates nor futures rates need be unbiased predictors of future spot rates in an efficient market. In each case the Lucas model allows a conditional covariance, which is related to the riskiness of an asset, to separate these prices from the expected value of the spot exchange rate. Since this conditional covariance can vary through time, the nature of the bias in forward rates and futures prices can be time varying.

Section 3 begins the discussion of the empirical analyses with an examination of the time-series properties of spot rates. Autocorrelations and empirical distributions of exchange rates are discussed. Then, tests of the unbiasedness of forward rates as predictors of future spot rates are examined. The relative merits of various strategies for testing the restrictions that such a hypothesis places on the data are discussed. This section incorporates a discussion of Hansen's [96] Generalized Method of Moments which is often a computationally convenient way of testing restrictions that

---

[1] Adler and Dumas [2] survey the development of theories of international asset pricing from a finance viewpoint. Their modeling strategy follows Solnik [198], Kouri [123], Hodrick [105], and Stulz [202] in the use of stochastic calculus to address issues in international asset pricing. Stulz [203] provides an introduction to mean variance optimization models in international asset pricing, while Stulz [204] extends his earlier analysis to incorporate money explicitly. Branson and Henderson [24] present a complete derivation of macroeconomic portfolio balance models, derive many results from the literature as well as present complementary microeconomic foundations of these models from the stochastic calculus models of international finance.

arise in rational expectations models. The major conclusion of Section 3 is that very strong evidence exists against the hypothesis that forward exchange rates, of any maturity from one day, to one week, to one or three months, are unbiased predictors of future spot rates.

Section 4 explores several alternative interpretations of the rejection of the unbiasedness hypothesis. Since the statistical arguments that are used to reject unbiasedness are invariably based on asymptotic distribution theory, there is always the possibility that the sample moments of the data are poor reflections of their asymptotic counterparts. While this argument applies with equal force to virtually all modern time series analysis, it may be that the types of government policies and other exogenous processes that determine exchange rates make this problem particularly strikingly manifest in these studies. Such a possibility probably deserves more study than it has received.

A second interpretation of the rejection of the unbiasedness hypothesis relies on Fama's [59] ingenious decomposition argument. By viewing the forward premium as the sum of two unobservable components, the expected rate of depreciation plus a normalized risk premium, and by considering the algebra of least squares, Fama demonstrated that rejection of the unbiasedness hypothesis in most instances implies that risk premiums are more variable than expected rates of depreciation and that the two covary negatively. While Fama found the results troublesome and potentially at variance with theory, the results of Hodrick and Srivastava [108] indicate that negative covariation per se is not inconsistent with the Lucas model. The magnitude of variability in the risk premiums remains an open question.

Section 4 continues with a discussion of the profitability of various trading strategies and contentions that the forward exchange market is inefficient. The work of Hodrick and Srivastava [106] questions whether Bilson's [14] trading strategy produces expected profits that are too good to be consistent with risk aversion. Assessment of the profitability of trading rules is complicated by the necessity, documented in Section 3, of allowing the conditional variance of forward rate forecast errors to vary over time. This problem also plagues analysis of the profitability of the interesting filter rule studies of Dooley and Shafer [44, 45] and Sweeney [207] that are addressed in this section.

A third alternative explanation of the rejection of the unbiasedness hypothesis involves the modeling of time varying risk premiums. Section 5 explores two complementary approaches to this subject. First, models that utilize only financial data are explored. This section includes a discussion of the latent variable model of Hansen and Hodrick [98] with its extensions by Hodrick and Srivastava [106], Campbell and Clarida [29], and Giovannini and Jorion [86].

The presence of conditional heteroscedasticity in forward rate forecast errors prompted Domowitz and Hakkio [42] to develop a model of the risk premium based on an autoregressive conditional heteroscedasticity framework. This is also discussed in Section 5.

Korajczyk [122] noted that the variability of risk premiums in theory can be related to variations in expected real interest rates. He developed and tested this approach which is discussed in Section 5. His empirical work also utilized bootstrap and Monte Carlo simulations in addition to the asymptotic distribution theory, and potential problems with this approach are reviewed.

Most of the econometric models of risk premiums that have involved market fundamentals, such as asset stocks, and not just financial data, have been of the two-period mean-variance type. Section 5 considers the papers of Frankel [67] and Frankel and Engel [69] in detail. The inability of this model to explain the data suggests that sterilized intervention in foreign exchange markets is unlikely to be successful, and Section 5 also surveys the literature relating to intervention and risk premiums.

The most direct test to date of the Lucas model was conducted by Mark [148]. He follows Hansen and Singleton [100, 101, 102] in adopting expenditures on nondurables and nondurables plus services as his measures of consumption. His estimation results and some potential problems with the approach are discussed in Section 5.

Section 6 explores the relation between forecasts of exchange rates and risk premiums. Frankel and Froot [70] and Froot and Frankel [79] have used available survey data of experts to attempt to quantify expected rates of depreciation and risk premiums. Unfortunately, their data series are relatively short, but the results are quite provocative and may be indicative of substantial non-stationarities in the data. Wolff [217] adopts a Kalman filtering approach to measuring risk premiums. He finds that the best forecasting models have highly autocorrelated risk premiums and large innovation variances of exchange rates relative to innovations in risk premiums. In no case does one of his models have a smaller mean-square error of its forecasts than the assumption that the exchange rate follows a random walk.

Section 7 examines the limited empirical work on futures prices of foreign exchange. The first issue addressed is whether there is any meaningful economic difference between futures prices and forward prices for the same terminal date in the future. Cornell and Reinganum [33] examined this issue and concluded that the potential difference that was derived by Cox, Ingersoll and Ross [34] does not manifest itself in the data. Consequently, if forward prices are biased predictors of future spot prices, futures prices must be biased predictors, as well, and by application of the law of iterated expectations, futures prices must be biased predictors of the next day's futures price. Hodrick and Srivastava [107] explored this hypothesis and their analysis is discussed. They are able to reject the hypothesis but find evidence of low variability of daily risk premiums. Consistency of daily and monthly evidence consequently requires highly serially correlated risk premiums. Some conflicting evidence on these issues is provided by McCurdy and Morgan [154]. They are unable to reject the unbiasedness hypothesis with weekly data.

Feinstone [62] also argues that such serial correlation is absent from the data. Her study involves the analysis of intradaily price changes in the foreign currency futures market. While nonnormality of price changes and nonstationarity of distribution across different days complicates the analysis of serial correlation, Feinstone argues that the movement of future prices is best modeled with a serially uncorrelated compound Poisson-normal process. With such a process prices jump only at the arrival of new information. If a jump occurs, its size is drawn from a normal distribution while the number of jumps in a given interval of time is assumed to be drawn from an independent Poisson distribution.

Conclusions and a short discussion of outstanding issues are presented in Section 8.

2. ASSET PRICING THEORY

There are many papers that provide a theoretical asset-pricing relationship between forward foreign exchange rates and expected future spot rates (examples include Solnik [198], Grauer, Litzenberger, and Stehle [88], Kouri [123], Stockman [200], Fama and Farber [60], Roll and Solnik [182], Frenkel and Razin [76], Hodrick [105], Lucas [145], Stulz [202], and Svensson [206]). These international asset pricing models are typically developed from the maximizing behavior of a representative agent in an infinite horizon environment. An alternative approach is the macroeconomic portfolio balance models of Frankel [65], Dornbusch [47], and Henderson [104]. The predictions of the theories regarding the relation between forward exchange rates and expectations of future spot rates are roughly similar in that all models generally predict deviations of the forward rate from the expected future spot rate. Of course, the theories differ in their predictions of the nature of this deviation and what drives its movement over time.

The primary theory developed in this monograph is taken from Lucas [145] as presented in Hodrick and Srivastava [106]. The Lucas model is a complete, dynamic, two country, general equilibrium model that provides useful insights into the nature of risk premiums in asset markets. The model provides a derivation of risk premiums in the forward market and in the futures market. The generality of the model allows the pricing of contracts for any maturity. It is useful to present these theoretical ideas prior to a discussion of the empirical evidence because much of the empirical literature provides a distorted view of what is meant here by an efficient market. Since the Lucas model is very abstract, it is not testable directly, and later sections demonstrate how additional auxiliary assumptions can generate empirically testable hypotheses.

2.1. The Lucas model

Consider now the structure of the model. The world is postulated to consist of two countries that are labeled country 0 and country 1. Agents in the two countries have identical preferences over two consumption goods but different stochastic endowments of the two goods. In period t, each citizen of country 0 is endowed with $2X_t$

units of commodity X and nothing of commodity Y, and each
citizen of country 1 is endowed with $2Y_t$ units of commodity Y and
nothing of commodity X. The representative agent's preferences are

$$E_0\left\{\sum_{t=0}^{\infty} \beta^t U(X_{it}, Y_{it})\right\} \tag{2.1}$$

where β is the common discount factor, $0 < \beta < 1$, and X_{it} and Y_{it}
are the representative agent's consumption of good X and good Y,
respectively, in country i in period t. The function U is assumed to
be bounded, continuously differentiable, increasing in both arguments, and strictly concave. The mathematical expectation operator
conditional on information available at time t is denoted $E_t(\cdot)$.

The current real state of the system is assumed to be completely
described by the realizations of X_t and Y_t. The bivariate process
(X_t, Y_t) is assumed to be drawn from a unique stationary Markov
distribution $F(X_t, Y_t \mid X_{t-1}, Y_{t-1})$ that gives the probability of X_t and
Y_t conditional on the realizations of X_{t-1} and Y_{t-1}.[2]

The equilibrium studied by Lucas is a stationary one in which
agents of the two countries pool risks in the available securities
markets that are assumed to be complete in the sense of Arrow [9]
and Debreu [39]. The complete pooling of risks allows each
representative agent to consume half of the endowment of each
good. In such an equilibrium, the relative price of Y in terms of X,
denoted P_t^y, depends only on the ratio of the marginal utility of Y to
the marginal utility of X,

$$P_t^y = U_t^y / U_t^x \tag{2.2}$$

where $U_t^y \equiv \partial U(X_{it}, Y_{it})/\partial Y_{it}$ and $U_t^x \equiv \partial U(X_{it}, Y_{it})/\partial X_{it}$.

Monies are introduced into the model in a very elementary way.[3]

---

[2] This is one of the simplest possible setups in which a formal equilibrium model
can be discussed. Obviously, more serial correlation of the real state, explicit
production and investment, and other features of the real world could be introduced
at a cost of increased complexity of the model.

[3] See Svensson [206] for an analysis that changes the timing of the receipt of
information and the occurrence of trading in markets for goods and assets. The
particular choice of timing used here has some unfortunate implications relative to
more complex models such as Svensson's. These implications are noted where
differences are important.

It is assumed that agents are required to purchase the endowment of a country only with the money of that country. At the beginning of a period, agents from both countries meet in a centralized market place, bringing securities and currency holdings previously accumulated, and engage in perfectly competitive securities trading. Before trading begins, the current period real state and any monetary innovations are assumed to be known to all traders.[4] Let M_t be the period t per capita quantity of the money of country 0, which is called dollars, and let N_t be the period t per capita quantity of the money of country 1, which is called pounds. Since all uncertainty about period t is revealed prior to trade in goods and securities, the purchasing powers of the dollar, π_t^m, and the pound, π_t^n, just depend on the ratio of $2X_t$ to M_t and $2Y_t$ to N_t, respectively:

$$\pi_t^m = 2X_t/M_t \tag{2.3}$$

$$\pi_t^n = 2Y_t/N_t, \tag{2.4}$$

where π_t^m is the good X price of the dollar and π_t^n is the good Y price of the pound.

There is also nominal uncertainty in the world that contributes to the uncertainty of the purchasing powers of the two monies. Monies are assumed to evolve over time as the realizations from a known Markov process. In each period t there is a lump sum dollar transfer of $w_{0t}M_{t-1}$ to agents of country 0 and a lump sum pound transfer of $w_{1t}N_{t-1}$ to agents of country 1. The transition function for the two monies is also a known Markov process $K[w_{t+1} \mid w_t, F(X_{t+1}, Y_{t+1} \mid X_t, Y_t)]$ where $w_t = (w_{0t}, w_{1t})$ is a vector of the stochastic growth rates of the two monies between $t-1$ and t. The function K is allowed to depend in a known way on the probability of the future real state given by F.

Given the relative price in (2.2) and the purchasing powers of the monies in (2.3) and (2.4), the equilibrium spot exchange rate is

---

[4] Lucas [144] notes that it is convenient to think of each trader as a two-person household consisting of a seller and a buyer. The seller supplies endowments to strangers thereby accumulating money balances during a period while the buyer purchases goods from other strangers thereby depleting money balances during the period. There is no possibility of intraperiod communication between the two parties. While this descriptive story rationalizes the cash-in-advance constraint, it plays no formal role in the analysis.

given by an arbitrage argument:

$$S_t = \pi_t^n P_t^y / \pi_t^{m}.^5 \tag{2.5}$$

The equilibrium prices of assets in this world can be derived in a style similar to other real intertemporal asset pricing models such as Rubinstein [186], Lucas [143], Brock [26], and others. The equilibrium price of an asset is determined by equating the marginal utility foregone from the purchase of the asset to the conditional expectation of the discounted marginal utility of the return from holding the asset. The conditional expectation is taken with respect to the distribution functions F and K.

Consider an arbitrary asset i with a dollar price V_t^i and a time $t + 1$ payoff of $V_{t+1}^i + D_{t+1}^i$ where D_{t+1}^i is either a dividend or coupon payment at time $t + 1$. The time $t + 1$ payment may be either deterministic or stochastic when viewed from time t. The foregone marginal utility of investing in this asset is $V_t^i \pi_t^m U_t^x$. The expected marginal utility of the payoff is $E_t[\beta U_{t+1}^x \pi_{t+1}^m (V_{t+1}^i + D_{t+1}^i)]$. It depends on the uncertain purchasing power of the dollar at time $t + 1$, on the uncertain marginal utility of commodity X at that time, and on the possibly uncertain payoff on the asset. Equating the marginal benefit to the marginal cost produces an expression that must be satisfied by all equilibrium dollar returns. The price of the asset adjusts to provide the appropriate expected return. A similar expression must hold for returns between time t and $t + k$:

$$E_t(Q_{t+k,k}^m R_{t+k,k}^i) = 1 \tag{2.6}$$

where $R_{t+k,k}^i \equiv (V_{t+k}^i + D_{t+k}^i)/V_t^i$ is a nominal dollar return from

---

[5] A weakness of this version of the Lucas model is that the spot exchange rate depends only on time t information. This occurs because of the timing of the receipt of information about period t and the opportunity to trade in the markets for assets and goods. When all information relevant for determining consumption is known prior to trade in money markets, the velocity of circulation is unity and the demand for money does not depend on interest rates. Changing the timing as in Svensson [206], or postulating cash goods and credit goods as in Lucas and Stokey [147] and Lucas [146] introduces a liquidity preference motive to money demand that produces a forward-looking component to the exchange rate. Since we are not concerned here explicitly with models of the determination of spot rates, the simplicity of (2.5) is not a problem.

investing at t with a payoff at $t + k$ and where

$$Q^m_{t+k,k} \equiv (\beta^k U^x_{t+k} \pi^m_{t+k} / U^x_t \pi^m_t). \qquad (2.7)$$

The expression in (2.7) defines the intertemporal marginal rate of substitution of dollars between time t and time $t + k$. Notice that $Q^m_{t+k,k}$ is dependent upon realizations of variables at time $t + k$. Hence, it is a random variable from the perspective of time t.

The intertemporal marginal rate of substitution of money is an index that weights the change in the purchasing power of a money between two time periods by the intertemporal marginal rate of substitution of goods between those time periods. The intertemporal marginal rate of substitution of good X between t and $t + k$ is $\beta^k U^x_{t+k} / U^x_t$. As Richard and Sundaresan [174] note, the expression $Q^m_{t+k,k}$ in conjunction with the conditional expectation operator can be thought of as a present value operator that provides dollar present values at time t of either stochastic or deterministic dollar payoffs at time $t + k$. In some of the discussion that follows it will be necessary to have two subscripts associated with a random variable such as a return or the intertemporal marginal rate of substitution. The first subscript indicates when the variable enters the information set of agents while the second denotes the amount of adjustment into the past or the future that is necessary to describe the variable correctly. Typically, if the return is one period, the second subscript will not be used.

In Hansen and Hodrick [98] and Hodrick and Srivastava [106] the first order condition of the representative agent (2.6) is used to derive testable restrictions relating forward exchange rates and expected future spot rates. These ideas were extended in Hodrick and Srivastava [107] to an examination of futures prices of foreign currency. These derivations are facilitated by a discussion of discount bill prices which are the dollar price at time t of a claim to one dollar at time $t + 1$ and the pound price at time t of the claim to one pound at time $t + 1$. Let B^m_t and B^n_t be the discount bill prices of dollars and pounds, respectively.

A claim to one dollar delivered with certainty at time $t + 1$ costs B^m_t dollars at time t. Applying (2.6) to this asset gives

$$B^m_t = E_t(Q^m_{t+1}) \qquad (2.8)$$

since the return is $1/B^m_t$ in this case, and B^m_t is in the time t

information set. A similar derivation of the pound price of purchasing one pound at time t for delivery with certainty at time $t + 1$ introduces the intertemporal marginal rate of substitution of pounds. If an investor sacrifices B_t^n pounds today, the loss in terms of utility is $B_t^n \pi_t^n U_t^y$. The expected gain in terms of utility is $E_t(\beta U_{t+1}^y \pi_{t+1}^n)$ which depends on the uncertain purchasing power of the pound and on the uncertain marginal utility of commodity Y in the future. Equating marginal cost and expected marginal benefit gives

$$B_t^n = E_t(Q_{t+1}^n) \tag{2.9}$$

where

$$Q_{t+k,k}^n \equiv (\beta^k U_{t+k}^y \pi_{t+k}^n / U_t^y \pi_t^n). \tag{2.10}$$

The expression in (2.10) defines the intertemporal marginal rate of substitution of the pound. Notice that $Q_{t+k,k}^n$ is also a random variable from the perspective of time t.

2.2. Forward prices of foreign exchange

In order to determine the nature of the risk premium in the forward foreign exchange market, the forward price must be derived. Let $G_{t,k}$ be the forward price of pounds in terms of dollars contracted at time t for delivery at time $t + k$. Consider the one-period forward price, G_t. If there is no default risk on nominal investments in bills denominated in dollars or pounds and no default risk on the forward contracts to purchase one currency for another at the known price, G_t, then investors must be indifferent between investing in the dollar denominated bill or investing the same amount of dollars in the pound denominated bill and selling the known pound proceeds in the forward market. Equality of these two dollar returns is a statement of interest rate parity,

$$(1/B_t^m) = (1/B_t^n)G_t/S_t. \tag{2.11}$$

If these transactions are truly risk free, this equation holds as an arbitrage condition. Studies by Frenkel and Levich [73, 74] and the analysis of McCormick [152] indicate that these markets are indeed well arbitraged with contemporaneous realizations of the variables in (2.11) falling within a neutral band induced by the small transactions costs in these markets.[6]

Equation (2.11) is one way to consider the determination of forward exchange rates. An alternative derivation relates the forward rate to the expected future spot rate. If one buys foreign exchange at the forward price $G_{t,k}$, the time $t+k$ payoff is $S_{t+k} - G_{t,k}$. Cox, Ingersoll and Ross [34] use this idea in developing an arbitrage strategy to price a forward contract. Let $R_{t,k}$ be the known dollar return from investing one dollar at time t in a risk-free discount bill whose payoff occurs at time $t+k$. Then, Cox, Ingersoll and Ross [34] demonstrate in their Proposition 1 that the forward exchange rate, $G_{t,k}$, is the present value of $S_{t+k}R_{t,k}$. The proof is straightforward. Consider an investment strategy of placing $G_{t,k}$ dollars in the k-period bill at time t and buying $R_{t,k}$ forward contracts to purchase foreign exchange. The time $t+k$ profit on the strategy is

$$G_{t,k}R_{t,k} + (S_{t+k} - G_{t,k})R_{t,k} = S_{t+k}R_{t,k}. \qquad (2.12)$$

Since the initial investment is $G_{t,k}$ dollars and the time $t+k$ payoff is $S_{t+k}R_{t,k}$, in equilibrium, arbitrage requires that $G_{t,k}$ be equal to the present value of $S_{t+k}R_{t,k}$. As noted above, (2.6) can be used to take present values of dollar payoffs. Hence,

$$G_{t,k} = E_t(Q^m_{t+k,k}S_{t+k}R_{t,k}). \qquad (2.13)$$

Relating the forward rate to the expected future spot rate requires some manipulation of the right-hand-side of (2.13). The definition of the conditional covariance between two random variables Z^1_{t+k} and Z^2_{t+k} is

$$C_t(Z^1_{t+k}; Z^2_{t+k}) = E_t(Z^1_{t+k}Z^2_{t+k}) - E_t(Z^1_{t+k})E_t(Z^2_{t+k}). \qquad (2.14)$$

Hence, (2.13) can be rewritten using (2.14) as

$$G_{t,k} = E_t(Q^m_{t+k,k}R_{t,k})E_t(S_{t+k}) + C_t(Q^m_{t+k,k}R_{t,k}: S_{t+k}) \qquad (2.15)$$

or

$$G_{t,k} = E_t(S_{t+k}) + C_t(Q^m_{t+k,k}R_{t,k}; S_{t+k}) \qquad (2.16)$$

---

[6] Dooley and Isard [43] examine the influence of explicit capital controls and what Aliber [7] called political risk or the threat of capital controls in determining the differential between interest rates on Deutsche mark denominated assets in German money markets versus Deutsche mark denominated assets outside of Germany.

where (2.16) follows from (2.6) because $R_{t,k}$ is a legitimate dollar return.

The forward exchange rate in theory is not an unbiased predictor of the future spot rate. The source of the bias is the conditional covariance in (2.16) which can be thought of as a risk premium. To see this, perform the same covariance decomposition on (2.6) giving

$$E_t(Q^m_{t+k,k})E_t(R^i_{t+k,k}) + C_t(Q^m_{t+k,k}; R^i_{t+k,k}) = 1. \qquad (2.17)$$

Multiply both sides of (2.17) by $R_{t,k}$ and recognize that $R_{t,k}E_t(Q^m_{t+k,k}) = 1$, since $R_{t,k}$ is in the time t information set. After rearrangement, (2.17) becomes

$$E_t(R^i_{t+k,k}) = R_{t,k} - C_t(Q^m_{t+k,k}R_{t,k}; R^i_{t+k,k}). \qquad (2.18)$$

The expected return between t and $t + k$ on the ith uncertain asset is equal to the known k-period return minus a conditional covariance between the ith return and the product of $R_{t,k}$ and the k-period intertemporal marginal rate of substitution of money. This conditional covariance is a risk premium which separates the expected return from the certain nominal return.

Rearrangement of (2.16) to conform to (2.18) gives

$$E_t(S_{t+k}) - G_{t,k} = -C_t(Q^m_{t+k,k}R_{t,k}; S_{t+k}). \qquad (2.19)$$

The left-hand side of (2.19) is the expected profit on a long position in the forward market, that is on a forward market purchase of foreign currency, while the right-hand side is the risk premium. No risk-free return enters (2.19) as it does in (2.18) because no resources must be sacrificed when making a forward contract.[7]

An alternative representation of the risk premium in (2.18) was proposed by Hansen and Richard [99] and was exploited empirically by Hansen and Hodrick [98]. Assume that agents can trade an asset whose return is

$$R^m_{t+k,k} = Q^m_{t+k,k} / E_t(Q^m_{t+k,k})^2. \qquad (2.20)$$

---

[7] This is not strictly correct in the sense that banks often require their forward foreign exchange customers to have approved lines of credit, and banks place additional limits on the amount that firms can draw on their lines of credit when the firms have outstanding forward contracts. The opportunity cost of the lost option to use the reduced line of credit could be incorporated into the asset pricing specification, but it is assumed to be zero.

Then, from (2.6) we see that this return is the minimum second moment return.[8] Hansen and Richard [99] establish that any return $R^b_{t+k,k}$ on the mean-variance frontier conditional on time t information satisfies

$$R^b_{t+k,k} = \omega_t R^m_{t+k,k} + (1 - \omega_t)R_{t,k} \qquad (2.21)$$

where ω_t is in the time t information set. The introduction of the benchmark return in (2.21) allows one to rewrite (2.18) in terms of a conditional capital asset pricing model:

$$E_t(R^i_{t+k,k}) = R_{t,k} + \beta^i_t E_t(R^b_{t+k,k} - R_{t,k}) \qquad (2.22)$$

where $\beta^i_t \equiv C_t(R^i_{t+k,k}; R^b_{t+k,k})/V_t(R^b_{t+k,k})$.[9]

Performing the analogous operations on (2.19) that translate (2.18) into (2.22) provides the representation of the risk premium that was exploited empirically in Hansen and Hodrick [98]:

$$E_t(S_{t+k} - G_{t,k})/S_t = \beta^G_t E_t(R^b_{t+k,k} - R_{t,k}), \qquad (2.23)$$

Where $\beta^G_t \equiv C_t[(S_{t+k} - G_{t,k})/S_t; \ R^b_{t+k,k}]/V_t(R^b_{t+k,k})$. Empirical discussion of this equation will be in Section 5.

---

[8] The proof is straightforward. Let $R^j_{t+k,k}$ be a legitimate return that satisfies (2.6). Write $R^j_{t+k,k} = Z/E_t(Q^m_{t+k,k})^2$. Consider placing 0.5 dollars into the two investments giving a return of $0.5(R^j_{t+k,k} + R^m_{t+k,k})$. Since this is a legitimate dollar return, it must satisfy (2.6). Therefore, $0.5E_t[(Q^m_{t+k,k}(R^j_{t+k,k} + R^m_{t+k,k})] = 1$. Substitution from the above definition of the jth return and from (2.20) gives $E_t(ZQ^m_{t+k,k}) = E_t(Q^m_{t+k,k})^2$. Since the second moment of the difference of the two returns must be positive,

$$E_t(R^j_{t+k,k} - R^m_{t+k,k})^2 = [E_t(Z^2) - 2E_t(ZQ^m_{t+k,k}) + E_t(Q^m_{t+k,k})^2]/[E_t(Q^m_{t+k,k})^2]^2,$$

and simplifying the numerator gives $[E_t(Z^2) - E_t(Q^m_{t+k,k})^2] > 0$. Therefore, $R^m_{t+k,k}$ as defined in (2.20) is the minimum conditional second moment return.

[9] To derive (2.22) from (2.18) first multiply and divide the covariance in (2.18) by

$$V_t(R^b_{t+k,k}) = E_t(R^b_{t+k,k})^2 - [E_t(R^b_{t+k,k})]^2 = \omega_t^2\{E_t(R^m_{t+k,k})^2 - [E_t(R^m_{t+k,k})]^2\}$$

$$= \omega_t^2\left\{\frac{1}{E_t(Q^m_{t+k,k})^2} - \frac{[E_t(Q^m_{t+k,k})]^2}{[E_t(Q^m_{t+k,k})^2]^2}\right\}.$$

Then, put $[\omega_t/E_t(Q^m_{t+k,k})^2]$ inside the covariance and move $R_{t,k}$ outside of the covariance. Both are legitimate operations because both variables are in the time t information set. The covariance becomes

$$C_t[\omega_t Q^m_{t+k,k}/E_t(Q^m_{t+k,k})^2; R^i_{t+k,k}] = C_t(R^b_{t+k,k}; R^i_{t+k,k})$$

while the term in brackets is $\omega_t[R_{t,k} - E_t(R^m_{t,k,k})]$ from (2.6). This latter term is $-(R^b_{t+k,k} - R_{t,k})$ from (2.21). Substituting gives (2.22).

2.3. Futures prices of foreign exchange

Cox, Ingersoll and Ross [34] also use an arbitrage argument to
determine the relation between futures prices of foreign exchange
and spot prices in the future. Let $F_{t,k}$ be the futures price of foreign
currency in terms of dollars at time t with a maturity date on the
contract of time $t + k$. Then, Cox, Ingersoll and Ross [34] establish
in their Proposition 2 that $F_{t,k}$ is the present value at time t of a
payoff of $S_{t+k} \prod_{i=t}^{t+k-1} R_i$ at time $t + k$, where present value is again
in reference to an appropriate asset pricing paradigm. Here the
operator $\prod_{i=t}^{t+k-1} Z_i$ indicates multiplication of Z_t by all Z_t's between
$t + 1$ and $t + k - 1$, and R_t is the one period nominal return between
t and $t + 1$ that is known at time t.

The proof of this proposition incorporates the institutional
feature of futures markets called marking to market. In futures
markets the clearinghouse of the futures exchange stands between
each buyer and seller of a futures contract. It takes no active
position in the market, but it interposes itself between all parties to
every transaction. In order to assure performance on a contract and
to keep all contracts essentially the same, the exchange requires
daily resettlement of profit and loss on a contract. If a person buys
foreign exchange in the futures market at $F_{t,k}$, the purchase is made
from someone who sold at that price. Both parties have contracts
with the clearinghouse. If on the following day the price rises (falls)
to $F_{t+1,k-1}$, the amount $F_{t+1,k-1} - F_{t,k}$ is credited to (debited from)
the account of the long party and debited from (credited to) the
account of the short party who sold the futures contract. This
intervening sequence of cash flows is one of the major differences
between futures contracts and forward contracts.

Now, consider the arbitrage argument in Cox, Ingersoll and Ross
[34]. At each time $j = t, t + 1, t + 2, \ldots, t + k - 1$ invest $F_{t,k}$ dollars
and the accumulated interest in one period risk-free bills with return
R_j. This gives a time $t + k$ payoff of $F_{t,k} \prod_{j=t}^{t+k-1} R_j$. Also, at each
point in time j, take a long position in $\prod_{i=t}^{j} R_i$ futures contracts at
price $F_{j,n}$ where $n \equiv t + k - j$. In each period after the first, liquidate
the contracts from the previous period to receive the per contract
profit or loss of $(F_{j+1,n-1} - F_{j,n})$ and invest the proceeds (which may
be negative) and the interest that accumulates in one period bills.

The time $t + k$ payoff from contracts entered into in the jth period is $\prod_{i=t}^{j} R_i(F_{j+1,n-1} - F_{j,n}) \prod_{i=j+1}^{t+k-1} R_i$. Since $F_{t+k,0} = S_{t+k}$ by arbitrage, the time $t + k$ payoff on the entire strategy is

$$F_{t,k} \prod_{j=t}^{t+k-1} R_j + \sum_{j=t}^{t+k-1} \left(\prod_{i=t}^{j} R_i \right)(F_{j+1,n-1} - F_{j,n})$$

$$\times \left(\prod_{i=j+1}^{t+k-1} R_i \right) = S_{t+k} \prod_{j=t}^{t+k-1} R_j. \quad (2.24)$$

Since the initial investment is $F_{t,k}$ dollars, the futures price must be set in equilibrium such that the value of sacrificing $F_{t,k}$ at time t is equal to the present value of $S_{t+k} \prod_{j=t}^{t+k-1} R_j$.

If the present value is taken as in the previous section, the futures price equation that corresponds to (2.13) is

$$F_{t,k} = E_t \left(Q^m_{t+k,k} S_{t+k} \prod_{i=t}^{t+k-1} R_i \right). \quad (2.25)$$

Since $\prod_{i=t}^{t+k-1} R_i$ is a legitimate return between t and $t + k$, it must satisfy (2.6) and the futures price equation that corresponds to (2.16) is

$$F_{t,k} = E_t(S_{t+k}) + C_t \left(Q^m_{t+k,k} \prod_{i=t}^{t+k-1} R_i; S_{t+k} \right). \quad (2.26)$$

The futures price is also a biased predictor of the future spot price, and in this case the bias is the conditional covariance on the right-hand side of (2.26) which can be considered a risk premium in the futures market.

Since the product of the one-period interest rates, $\prod_{i=t}^{t+k-1} R_i$, is a random variable from the perspective of time t, the right-hand side of (2.25) is not equal to the right-hand side of (2.13). Black [16] first noted that forward prices would equal futures prices if the product of the short rates were a deterministic function and hence equal to $R_{t,k}$ by arbitrage. The fact that the product of the short rates is generally stochastic implies that one cannot decompose (2.26) to produce a conditional capital asset pricing expression as was done in deriving (2.23).

Subtracting (2.25) from (2.13) gives the difference between the forward rate at time t for delivery at time $t + k$ and the futures price at time t on a contract that matures at time $t + k$:

$$F_{t,k} - G_{t,k} = E_t\left[Q^m_{t+k,k}S_{t+k}\left(\prod_{i=t}^{t+k-1} R_i - R_{t,k}\right)\right]. \qquad (2.27)$$

The right-hand side of (2.27) is the dollar present value of a random amount of foreign exchange equal to the product of the k short rates minus the return on a k-period bond.

This section has presented the derivation of forward and futures prices for foreign exchange in the context of an intertemporal asset pricing model. The key expressions (2.16) and (2.26) demonstrate that these speculative prices are generally theoretically biased predictors of future spot exchange rates in an efficient market. The next sections address the empirical literature that has attempted to examine and test these relations.

3. ECONOMETRIC TESTS OF THE EFFICIENCY HYPOTHESIS: AUTOCORRELATION AND UNBIASEDNESS

This section develops and discusses various empirical tests that seek to assess the efficiency of the foreign exchange market. At the outset it is important to note that as with other financial markets any test of market efficiency is a joint test of several composite hypotheses. Hence, it is impossible to develop a direct test of the hypothesis that the foreign exchange market is efficient. All that can be done is to present various statistical hypotheses regarding what one means by market efficiency and test these specifications by placing additional assumptions on the statistical properties of the data.[10] Rejection of the null hypothesis is consequently not necessarily identified with market inefficiency. Indeed, it is perhaps more appropriate to interpret the statistical analysis in a quasi-Bayesian way, as is argued by Leamer [127]. We have some priors about how the world works, and interaction with the data allows us to update our thinking. It also allows us to assess how well our priors explain

---

[10] See Fama [58] for a clear statement of these ideas as applied to returns in the stock market. The ideas of weak, semi-strong, and strong form efficiency that were discussed in Fama [57] are presented in terms of models of equilibrium expected returns. Fama writes [58, p. 168], "Formal tests require formal models, with their more or less unrealistic structuring of the world."

the data and to begin development of new theories that may do a better job in the future.

Levich [133] notes that part of the confusion surrounding tests of efficiency of foreign exchange markets is generated by an application to foreign exchange markets of ideas from the early finance literature on efficiency of stock markets. Fama [57] argued that an efficient market is a market where prices "fully reflect" all available information. In such a circumstance no investor or speculator can earn extraordinary profits by exploiting publically available information. This does not imply that equilibrium expected returns on assets are all the same, though, since assets may differ in their riskiness. Also, there is no necessary presumption that the equilibrium expected return on an asset is constant through time. These qualifications that expand the definition of an efficient market make testing the concept quite difficult. The ideas also generally imply that tests of efficiency in the forward foreign exchange market are necessarily joint tests of an equilibrium model of expected returns and rational processing of available information by investors. One specifies a model of equilibrium expected returns and an information set of investors, and one postulates that economic agents set asset prices to make expected returns on assets conform to the expected values predicted by the model.

A third problem that plagues statistical studies of efficient markets is that some statistical properties must be assumed for the time series used in the analysis. Typical assumptions include stationarity and ergodicity.[11] Virtually all rational expectations econometric techniques require that the sample moments from a large sample of data converge to the true moments of the population. Unfortunately, financial and economic data may require relatively large samples before we experience all of the possible events on which agents place prior probability. In this literature, the failure of sample moments to coincide with population moments due to the presence of an event that has not occurred tends to be called the 'peso problem' since it was first realized in considerations of the relation of forward rates to spot rates in the presence of an expected devaluation. A possible solution to this type of problem was proposed by Krasker [124], who considers one particular type

---

[11] See Rozanov [185] for a discussion of the formal meaning of these terms.

of solution, but his approach is not a solution to the general problem.[12]

Although the previous section demonstrated that forward rates and futures prices are theoretically not unbiased predictors of future spot rates, much of the empirical work in the area has been directed to investigation of this hypothesis. Before examining the evidence, it is important to consider one problem with unbiasedness as a characterization of the no-risk-premium hypothesis.

3.1. The Siegel paradox

After the beginning of the modern experience with flexible exchange rates in 1973, there was considerable controversy regarding how one ought to characterize efficiency in the forward foreign exchange market. In Siegel [195] the proposition that the level of the forward rate was equal to the expected value of the level of the future spot rate was shown to imply a contradiction. If the proposition were true for exchange rates quoted as Japanese yen per Swiss franc, it could not be true for exchange rates quoted as Swiss francs per Japanese yen. This is because Jensen's inequality requires that $E(1/Z)$ be greater than $1/E(Z)$ when Z has positive variance. What came to be called Siegel's paradox was resolved by Roper [183] and Boyer [23] who demonstrated that risk neutrality and maximization of a utility function specified in real terms did not imply unbiasedness in the levels of the variables regardless of the way they are quoted.

In terms of the model developed in the previous section, risk neutrality implies a linear utility function and constant marginal utility. Consequently, the intertemporal marginal rate of substitution of money does not depend on consumption, and it simplifies to the ratio of the purchasing powers of the money times the discount factor. As a result, the expression in (2.13) is

$$G_{t,k} = E_t(\beta^k S_{t+k} \pi_{t+k}^m R_{t,k}/\pi_t^m) \qquad (3.1)$$

or using covariance decomposition,

$$G_{t,k} = E(S_{t+k}) + C_t(S_{t+k}; \beta^k \pi_{t+k}^m R_{t,k}/\pi_t^m). \qquad (3.2)$$

---

[12] Formal Bayesian analysis of these issues is also extremely difficult because the analyst must specify all possible events.

This idea has been addressed by McCulloch [153], Krugman [125], Stockman [200], Frankel [65], and Frenkel and Razin [76], and it has been tested empirically by Engel [54], which is discussed below. Even though agents are risk neutral, there still must be a nominal expected profit or loss on the forward contract in order to keep expected real profits equal to zero. Stockman [200] demonstrates that a Jensen's inequality term can always be carried separately in equations like (2.13), and he argues that McCulloch's [153] argument allows one to ignore it empirically. Such arguments typically rely on models in which purchasing power parity holds. Since the evidence against purchasing power parity is so strong, this argument may have little relevance. The reason deviations from unbiasedness are viewed here as evidence of a risk premium is the strong *a priori* conviction that I have, given observations on average returns across assets. Reconciliation of these observations with market efficiency requires risk aversion of economic agents. Hence, if the representative agent in equity and bond markets is risk averse, the marginal agent also will likely be risk averse in foreign exchange markets.

3.2. Autocorrelation based tests

Another area of considerable controversy in the early empirical literature on the efficiency of the foreign exchange market concerns the time series properties of spot exchange rates. Section 2 of this monograph describes a general equilibrium model in which the spot exchange rate follows the time series process induced by the exogenous Markov processes for money supplies and real incomes.[13] Clearly, the exchange rate need not follow a random walk in that model, yet the first empirical studies were concerned with this issue and tended to identify failure of a random walk with market inefficiency.

Levich [128] notes that equilibrium asset market rates of return

---

[13] Samuelson's [188] argument that perfectly anticipated asset prices fluctuate randomly was derived in an environment of risk neutrality and only for the sequence of futures prices that forecast a common date in the future. It does not imply that stock prices or exchange rates follow a random walk. See Lucas [143] for a general equilibrium asset pricing model with expected returns on equities that fluctuate through time.

need not be constant nor should exchange rates necessarily follow a random walk or a random walk with constant drift. He therefore argues that time series analysis of exchange rates serves primarily as a descriptive technique, which is how these studies are viewed here.

Early empirical work in this area is by Poole [172], who analyzes the serial correlation of rates of change of exchange rates for nine currencies from the period 1919 to a range of dates from 1924 to 1928. The exchange rates are the US dollar values of the currencies of Argentina, Belgium, Canada, France, Italy, Japan, Norway, Sweden and the United Kingdom. He also analyzes the US dollar-Canadian dollar exchange rate from 1950 to 1962. Poole finds significant first order serial correlation in many cases, and he also finds evidence of potentially large profits from simple trading rules.

In contrast, other early studies find little evidence of serial correlation. Levich [129] finds the serial dependence of the Canadian case in the 1950s to be associated only with the first and last years of their managed float. Giddy and Dufey [85] use Box–Jenkins time series techniques to analyze the 1920s. Their investigation indicates that while the simple random walk does not appear to be an adequate description of the data within the sample, the forecasts from simple Box–Jenkins models have larger mean squared errors out of sample than does the random walk. The source of the instability in the apparent serial correlation has not been documented. One possibility is that it is spuriously due to a failure of the exchange rate time series to satisfy the requisite assumptions necessary for the application of statistical tests. These assumptions are discussed below.

Cornell and Dietrich [32] and Dooley and Shafer [44] analyze the autocorrelations of daily rates of change of exchange rates beginning their samples with the breakdown of the Bretton Woods system in early 1973. Cornell and Dietrich [32] analyze data from March 1973 to September 1975 for US dollar values of the Canadian dollar, the Swiss franc, the Dutch guilder, the Deutsche mark, the UK pound, and the Japanese yen. They examine the autocorrelations from lag one to lag eight. None is larger in absolute value than 0.10, and most are substantially smaller than that. Only four of the 48 coefficients are statistically significant at the 95 percent confidence level. Their results are not substantially altered by substituting the rate of change of the actual exchange rate minus a proxy for

TABLE I

Autocorrelations of daily excess rates of change of exchange rates from
Dooley and Shafer [45]

Currency	Q-Statistics			
	Sample 1	Sample 2	Sample 3	Sample 4
Belgian franc	35.86	40.48[a]	52.44[a]	21.86
Canadian dollar	37.49	48.68[a]	46.14[a]	40.56[a]
French franc	35.29	40.27[a]	46.41[a]	24.45
Deutsche mark	32.35	45.38[a]	32.71	37.24
Italian lira	33.46	89.50[a]	206.60[a]	38.60[a]
Japanese yen	21.59	48.85[a]	52.03[a]	36.92
Dutch guilder	45.65[a]	43.40[a]	43.10[a]	32.00
Swiss franc	28.83	33.12	26.89	45.75[a]
UK pound	39.21[a]	44.01[a]	27.09	32.00

Notes: The Q-statistic is $N \sum_{i=1}^{20} \hat{\rho}_i^2$ where N is the sample size and $\hat{\rho}_i$ is the estimated autocorrelation coefficient at lag i. The statistic is distributed as a chi-square variable with 20 degrees of freedom. The one percent critical level of 37.60 is indicated with a superscript a. Sample 1 is 13 March 1973 to 5 September 1975. Sample 2 is 8 September 1975 to 6 November 1981. Samples 3 and 4 correspond to the first and second halves of Sample 2, respectively.

its predicted one day rate of change that was generated from the 30-day or 90-day forward premium.

In contrast to these results which focus on low order autocorrelations, Dooley and Shafer [45] report statistical tests of the hypothesis that the first twenty autocorrelations of daily excess rates of change of exchange rates are zero. Table I is taken from their Table 3-2. Dooley and Shafer [45] subtract the overnight Eurocurrency interest differential (or the best available alternative) between the US dollar and the other currencies from the daily rate of change of the dollar denominated exchange rates. The exchange rates are the dollar prices of the currencies of Belgium, Canada, France, Germany, Italy, Japan, the Netherlands, Switzerland, and the United Kingdom. Since the overnight interest differential can be equated to a 1 day forward premium as in (2.11), Dooley and Shafer's test can also be interpreted as a test that the 1 day forward rate is an unbiased predictor of the spot rate one day in the future. Table I indicates that there is considerable evidence against this hypothesis especially in the sample of data from 1975 to 1981. Most of the Q-statistics are well above the critical level of a chi-square with 20 degrees of freedom associated with the 0.01 marginal level of significance.

3.2a. The Distribution of changes in exchange rates

Cornell and Dietrich [32] note an important qualification to the above analysis. The tests of serial correlation are based on the assumption that innovations in the series are independently and identically distributed normal random variables. Nonnormality could bias the Q-statistics against the null hypothesis. They provide some graphical evidence but do no explicit tests of the normality of the series. Their graphical evidence indicates that, as with other financial return data, such as the stock market data analyzed by Fama [58] or the commodity futures data analyzed by Dusak [50], daily rates of change of exchange rates tend to have distributions with fatter tails than the normal distribution.

There appears to be a general consensus that changes in the natural logarithms of exchange rates over short-term intervals such as a day or a week have a distribution that has fatter tails than a normal distribution. There is less of a consensus over the source of this leptokurtosis.

Dooley and Shafer [44] present evidence against the hypothesis that daily exchange rate changes are normally distributed. In a similar vein, Westerfield [213] finds support for the hypothesis that one week rates of return on spot and forward contracts are not normally distributed, and she argues they are more likely drawn from a stable Paretian distribution for which variance is undefined. Rogalski and Vinso [177] reexamine Westerfield's data and conclude that a Student-t distribution, which has fatter tails than the normal but for which the variance is defined, is a more appropriate characterization of the data. This is consistent with an argument in Dooley and Shafer [44]. If daily changes are drawn from a stable distribution with undefined variance, then cumulating these changes into weekly and monthly changes would provide no indication that longer term changes are normally distributed. If, on the other hand, the daily changes are independent draws from any finite variance distribution, the central limit theorem assures that changes over longer time intervals will converge to a normal distribution. Their analysis of longer differencing intervals of the logarithms of exchange rates suggests that these intervals are closer to the normal distribution than are the daily changes.

McFarland, Pettit and Sung [155] examine daily changes in the natural logarithms of the US dollar values of four major currencies, the UK pound, the Deutsche mark, the Japanese yen, and the Swiss franc as well as three minor currencies, the Australian dollar, the Spanish peseta, and the Swedish krona. They conduct tests to determine whether the distribution differs by day of the week. They conclude that different days are characterized by different stable distributions even though the estimates of the characteristic exponent appears to increase as the days are added together.

Such a conclusion is inconsistent with the results of Friedman and Vandersteel [77]. They examine 1640 daily observations from 1 June 1973 to 14 September 1979 on changes in natural logarithms of the US dollar values of nine currencies: the Deutsche mark, the Swiss franc, the British pound, the Japanese yen, the Dutch guilder, the French franc, the Canadian dollar, the Belgian franc, and the Italian lira. They find that cumulating the changes does cause the estimate of the characteristic exponent to increase, but not conclusively toward two, its value under normality. Their more telling experiment occurs when they cumulate randomly permuted versions of each series. The stability property of the stable Paretian family is not supported by this experiment. Friedman and Vandersteel [77] therefore argue that the increase in the estimated value of the exponent is consistent with daily changes that are drawn from independent normal distributions whose parameters are time dependent.

Boothe and Glassman [20] and Hsieh [112] have recently examined ten years of daily data. The first authors examine not only daily and weekly changes but monthly and quarterly changes as well. They conclude that while normality is easily rejected at the daily and weekly level, it is never rejected with quarterly differences for US dollar values of the Canadian dollar, the UK pound, the Deutsche mark, and the Japanese yen.

Hsieh [112] divides his sample into two halves to examine the hypothesis of independently and identically distributed observations. By partitioning the event space into k mutually exclusive and exhaustive regions, he examines the equality of two multinomial distributions with a chi-square statistic. The hypothesis is strongly rejected as is expected given the previous results. The interesting part of the analysis occurs when the data are standardized by subtracting the estimated monthly mean and dividing by the estimated monthly standard deviation. Recomputation of the chi-square tests indicates that the hypothesis of independently and identically distributed data cannot be rejected at even the ten percent marginal level of significance for the US dollar values of the Canadian dollar, the Deutsche mark, and the Swiss franc while it can be rejected at the one percent level for the British pound and the Japanese yen.

All of the studies described above use data strictly from the flexible rate period while Farber, Roll, and Solnik [61] also examined data from the 1950's and 1960's. Their findings were for the US dollar values of 17 currencies, and they examined raw percentage changes in exchange rates as well as percentage changes in exchange rates adjusted for nominal interest differentials. Their findings for flexible periods confirm the findings discussed above, but the distributions under the fixed rate period are quite different. They find (p. 245), "the fixed rate period was characterized by larger probabilities of *extreme* changes." As they note, the regimes of exchange rates that have been classified as fixed have hardly been that.

Several of the early studies described above share a common desire to model the rate of change of the exchange rate as if it were drawn from a single unchanging distribution. More recent work has moved away from this idea. One theme of later sections of this monograph is that daily or even monthly exchange rate changes are probably not drawn from a single distribution, nor is it likely that they are drawn from a distribution with changing conditional mean and constant conditional variance. The presence of conditional heteroscedasticity also creates uncertainty about the interpretation of the test statistics reported in Table I.

Since autocorrelation tests are inconclusive and unpowerful in environments with changing conditional distributions, Cornell and Dietrich [32] and Dooley and Shafer [45] also report nonparametric runs tests of the hypothesis that exchange rate changes in excess of interest differentials are random. Of the six currencies in their study, Cornell and Dietrich report that only the Canadian dollar exhibits a significantly smaller number of runs than is expected, whereas Dooley and Shafer report that their entire sample has only two instances of runs in exchange rate changes that are longer than could have occurred by chance. Close examination of these events indicates that they were associated with substantial adjustments of international reserves at the central banks of Germany, Belgium, and Italy. This evidence suggests that new information is readily incorporated into the level of exchange rates making their changes near random walks.

3.3. Tests of unconditional unbiasedness

Some of the earliest empirical work on forward exchange rates as predictors of future spot rates examined the proposition that the mean forecast error is zero. Aliber [8], Cornell [31], Levich [128], Kohlhagen [121], Frankel [66] and Agmon and Amihud [4] examine the mean error or mean-squared error, and conclude that while forward rate forecast errors are large, they are not unconditionally biased. More recent evidence by Korajczyk [122] suggests that forward rates may have unconditional bias during his sample.

The next section begins the exploration of the more interesting question of conditional bias. If the asset pricing theory of Section 2 is correct, the risk premium separating forward rates from expected future spot rates can vary through time and no unconditional bias need be found, yet at each point in time the forward rate can differ from the expected future spot rate.

3.4. Derivation of regression based tests of unbiasedness

Section 3.2 demonstrates that exchange rates appear to be highly unpredictable. If they were actually random walks, their changes would be completely unpredictable. This section examines whether regression tests confirm evidence of predictable changes in exchange rates.

Frenkel [71] provides an early regression based empirical investigation of the unbiasedness hypothesis. His empirical research relies on a specification of the model in natural logarithms:

$$\ln G_{t,k} = E_t(\ln S_{t+k}). \tag{3.3}$$

Hansen and Hodrick (1983) offer an alternative interpretation of such a specification. If the variables in (2.13) are log-normally distributed, with constant conditional second moments, (3.3) with an additional constant term characterizes the data even with risk averse investors.

If we equate the market's subjective expectation in (3.3) based on a common information set with the mathematical expectation based on an information set available to the econometrician, we make the assumption of rational expectations, and we can proceed empirically. The hypothesis of rational expectations implies that any variable Z_{t+k} may be written as

$$Z_{t+k} = E_t(Z_{t+k}) + \varepsilon^z_{t+k,k} \qquad (3.4)$$

where $\varepsilon^z_{t+k,k}$ is the innovation or unanticipated part of Z_{t+k} that could not be predicted with time t information. It has a mean of zero. These properties of $\varepsilon^z_{t+k,k}$ imply that the covariance between $\varepsilon^z_{t+k,k}$ and any variable in the time t information set is zero. Substituting from (3.4) into (3.3) and letting lower case letters represent the natural logarithms of their upper case counterparts gives

$$s_{t+k} = g_{t,k} + \varepsilon_{t+k,k}. \qquad (3.5)$$

This specification motivated Frenkel [71] and others to perform an ordinary least squares regression such as

$$s_{t+k} = \alpha + \beta g_{t,k} + \varepsilon_{t+k,k} \qquad (3.6)$$

to test $\alpha = 0$ and $\beta = 1$ as the null hypothesis.[14]

Several comments are in order regarding such a specification. First, since $\varepsilon_{t+k,k}$ is generated by new information that arrives between time t and time $t + k$, the residuals of (3.6) will be serially correlated even under the null hypothesis unless $k = 1$. Frenkel [71] and others, therefore, sample the data to produce a data set with nonoverlapping residuals. Hansen and Hodrick [97] demonstrate how to avoid sampling the data which reduces the degrees of freedom, and this topic is discussed below in detail.

Second, consistency of the parameter estimates is assured if $\varepsilon_{t+k,k}$

---

[14] Using a specification like (3.6), Longworth [139] fails to reject the null hypothesis $\alpha = 0$ and $\beta = 1$ while Frenkel [66] finds statistically significant α coefficients and β coefficients significantly less than one. Edwards [51], Frenkel [72], and Park [171] also report results with specifications like (3.6).

has a finite variance and $\sum_{t=1}^{T} g_{t,k}^2 \to \infty$ as $T \to \infty$, since $\varepsilon_{t+k,k}$ is orthogonal to all time t information. This condition is satisfied, almost surely, when $g_{t,k}$ has a finite autoregressive representation with roots inside, on, or outside the unit circle.

Testing the null hypothesis, though, requires an asymptotic distribution theory for the estimators. Since $g_{t,k}$ is not an exogenous variable, in the sense that knowledge of $g_{t+h,k}$ for $1 \le h < k$ would provide useful information about $\varepsilon_{t+k,k}$, it must be considered a stochastic regressor that is a predetermined variable at time t. In the context of the linear model (3.6), some of the weakest conditions on the moment matrix of the explanatory variables that allow derivation of an appropriate asymptotic distribution are discussed in Grenander [90]. He demonstrates that asymptotic normality of the coefficient vector precludes exponential growth of any variable. This rules out explosive or nonstationary time series processes. Borderline nonstationarity or a unit root is allowed only if the regressors are fixed or strictly exogenous.

3.5. Tests for unit roots

Given the importance of stationarity in determining the asymptotic distribution of the coefficient vector, Meese and Singleton [160] were led to test whether the univariate processes of the natural logarithms of spot and forward exchange rates contain unit roots. Their tests are based on the work of Fuller [80], Dickey and Fuller [41] and Hasza and Fuller [103]. Meese and Singleton [160] use weekly observations on spot and three month forward rates for the US dollar values of the Swiss franc, the Deutsche mark, and the Canadian dollar. They state [160, p. 1032] "These results suggest that $\ln S_t$ and $\ln F_t$ do not have stable univariate AR representations, even after removing a linear trend."

Wasserfallen and Kyburz [211] also find strong evidence of unit roots in their investigation of the Swiss franc values of the Deutsche mark, the French franc, the British pound, and the Italian lira. These results are consistent with the studies discussed above in Section 3.2. If the levels of the logarithms of exchange rates were stationary, the first differences would show significant serial correlation. Clearly, the time series properties of the data do not support the specification of an equation like (3.6).

3.6. A conjecture about regressions in levels of exchange rates

Since Frenkel [72] argues that the regression evidence from (3.6) is supportive of the null hypothesis of unbiasedness, it is desirable to investigate a potential explanation why use of this approach may produce evidence that is supportive of the null hypothesis even though, as will be demonstrated below, it can be strongly rejected in specifications that use data in a form more likely to satisfy the assumption of time series stationarity.[15] From (3.3), an appropriate time series specification of the null hypothesis of unbiasedness is

$$E_t(s_{t+k} - s_t) = g_{t,k} - s_t, \tag{3.7}$$

where the left-hand side of (3.7) is the expected rate of depreciation of one currency relative to another and the right-hand side is the logarithmic forward premium. An appropriate OLS regression specification of (3.7) with nonoverlapping data is

$$s_{t+1} - s_t = \alpha + \beta(g_t - s_t) + \varepsilon_{t+1}, \tag{3.8}$$

and secondary subscripts on g_t and ε_{t+1} have been suppressed.[16] In what follows, if the forward rate is for the same period as the sampling interval, no secondary subscript will be used. If $\beta \neq 1$, as is the case under an alternative hypothesis that the unbiasedness hypothesis does not characterize efficiency in the forward market, (3.8) can be written as

$$s_{t+1} = \alpha + \beta g_t + (1 - \beta)s_t + \varepsilon_{t+1}. \tag{3.9}$$

Consider the sample estimate of β in (3.9) that is produced by ordinary least squares estimation of s_{t+1} on a constant and g_t for a sample of size T:

$$\hat{\beta} = \sum_{t=1}^{T} (s_{t+1} - \bar{s})(g_t - \bar{g}) \Big/ \sum_{t=1}^{T} (g_t - \bar{g})^2 \tag{3.10}$$

---

[15] Robert Flood first suggested this argument to me as a potential reason why results from the specification (3.6) yield such strikingly different implications than those reported below for alternative specifications. Meese [157] presents the argument more technically in terms of cointegration.

[16] Tryon [209], Bilson [14], Longworth [139], and Fama [59] investigate this specification of the unbiasedness test. The results generally reject the unbiasedness hypothesis, and they are discussed in Sections 3.12 and 4.

where $\bar{s} = (1/T) \sum_{t=1}^{T} s_{t+1}$ and $\bar{g} = (1/T) \sum_{t=1}^{T} g_t$ are the sample means. Substituting from (3.9) into (3.10) and using the algebra of least squares gives

$$\hat{\beta} = \sum_{t=1}^{T} \{\beta(g_t - \bar{g})^2 + (1 - \beta)(s_t - \bar{s})(g_t - \bar{g})$$

$$+ \varepsilon_{t+1}(g_t - \bar{g})\} \Big/ \sum_{t=1}^{T} (g_t - \bar{g})^2. \tag{3.11}$$

When g_t and s_t are nonstationary, the sample variance, $(1/T) \sum_{t=1}^{T} (g_t - \bar{g})^2$, and the sample covariance, $(1/T) \sum_{t=1}^{T} (s_t - \bar{s})(g_t - \bar{g})$, do not converge to any population values. What is true in the actual data, though, is that s_t and g_t tend to move roughly together with common large unanticipated components. Hence, it is very likely that sample values of $\sum_{t=1}^{T} (g_t - \bar{g})^2$ and $\sum_{t=1}^{T} (s_t - \bar{s})(g_t - \bar{g})$ are quite close numerically. Since $\sum_{t=1}^{T} \varepsilon_{t+1}(g_t - \bar{g})$ is likely to be close to zero whether g_t is stationary or not because ε_{t+1} depends only on new information, the sample estimate of $\hat{\beta}$ in (3.11) is driven to 1 regardless of the value of the true β. Since this is precisely the value predicted by the unbiasedness hypothesis, it seems particularly important that estimation proceeds under a specification that does not have this potentially undesirable property. The studies that are discussed below are therefore ones that work with a representation that is more likely to satisfy the stationarity assumption.

3.7. Maximum likelihood estimation

In order to avoid the problem of discarding a relatively large part of a sample to acquire a nonoverlapping sample, Hansen and Hodrick [97] consider the general problem of estimating a k-step-ahead linear forecasting equation

$$E(y_{t+k} \mid \Phi_t) = x_t \beta \tag{3.12}$$

where x_t is an r-dimensional row vector of variables contained in the information set Φ_t, and β is an r-dimensional column vector of parameters. The empirical procedures they employ are special cases of Hansen's [96] Generalized Method of Moments.

One way to proceed in such a case, which is discussed in Hansen and Hodrick [97] but is not employed there because of computational infeasibility, is to estimate β and test the constraints implied by (3.12) in a maximum likelihood estimation. That is, assume that the joint (y_t, x_t) process is covariance stationary and multivariate normal and trace through the restrictions implied by (3.12) on the Wold decomposition of the series. A Wold decomposition of (y_t, x_t) can be written as

$$y_t = \sum_{i=0}^{\infty} \delta_i v_{t-i} + \sum_{i=0}^{\infty} \gamma_i w_{t-i} + d_{yt} \tag{3.13}$$

$$x_t' = \sum_{i=0}^{\infty} \alpha_i v_{t-i} + \sum_{i=0}^{\infty} \psi_i w_{t-i} + d_{xt} \tag{3.14}$$

where $v_t = y_t - E(y_t \mid y_{t-1}, y_{t-2}, \ldots, x_{t-1}, x_{t-2}, \ldots)$, $w_t = x_t' - E(x_t' \mid y_{t-1}, y_{t-2}, \ldots, x_{t-1}, x_{t-2}, \ldots)$, $\delta_0 = 1$, $\gamma_0 = 0$, $\alpha_0 = 0$, $\psi_0 = I$, the r-dimensional identity matrix, and d_{yt} and d_{xt} are the deterministic components of y_t and x_t, respectively, which are assumed to be zero. The processes v_t and w_t are the innovations in y_t and x_t. Using (3.13) the conditional prediction of y_{t+k} given the history of y_t and x_t is

$$E(y_{t+k} \mid y_t, y_{t-1}, \ldots, x_t, x_{t-1}, \ldots) = \sum_{i=k}^{\infty} \delta_i v_{t+k-1} + \sum_{i=k}^{\infty} \gamma_i w_{t+k-i}. \tag{3.15}$$

Also, using (3.14) the process $x_t\beta$ may be written as

$$x_t\beta = \sum_{i=0}^{\infty} \beta' \alpha_i v_{t-i} + \sum_{i=0}^{\infty} \beta' \psi_i w_{t-i}.$$

Under the null hypothesis, $E(y_{t+k} \mid y_t, y_{t-1}, \ldots, x_t, x_{t-1}, \ldots) = x_t\beta$, we obtain the cross equation restrictions:

$$\begin{aligned} \beta' \alpha_i &= \delta_{i+k} \\ \beta' \psi_i &= \gamma_{i+k}. \end{aligned} \qquad i = 0, 1, 2, \tag{3.16}$$

If y_t and other lagged y's enter the x_t vector, additional cross-equation restrictions arise.

Because the joint covariance properties of the (y_t, x_t) series are totally characterized by the δ_i's, the γ_i's, the α_i's, the ψ_i's, and the parameters of the covariance matrix of v_t and w_t, maximum likelihood estimation requires estimation of these parameters subject to the restrictions in (3.16). Computationally, a finite parameter vector ζ representing the fundamental parameters must be concurrently estimated with the parameter of interest, β. The likelihood function is highly nonlinear in the parameters ζ, and maximization requires iterative numerical techniques. In the problem considered by Hansen and Hodrick [97] such a procedure is computationally intractable since the forecasting interval is thirteen, that is the data consist of weekly observations on spot and three-month forward exchange rates.

Baillie, Lippens and McMahon [12] and Hakkio [93] demonstrate one way to translate the restrictions on the Wold representation in (3.16) into testable hypotheses on a vector autoregression. They set $y_t = s_t$ and $x_t\beta = g_t$ where $s_t = \ln S_t$ and $g_t = \ln G_{t,4}$. That is, they choose weekly data and the one-month forward rate. They approximate the infinite order parameterization of the bivariate Wold decomposition with a finite order vector autoregression as in

$$\begin{vmatrix} A(L) & B(L) \\ C(L) & D(L) \end{vmatrix} \begin{vmatrix} s_t \\ g_t \end{vmatrix} = \begin{vmatrix} v_t \\ w_t \end{vmatrix} \tag{3.17}$$

where

$$A(L) = 1 - \sum_{j=1}^{p} A_j L^j, \qquad B(L) = \sum_{j=1}^{p} B_j L^j,$$

$$C(L) = \sum_{j=1}^{p} C_j L^j, \qquad D(L) = 1 - \sum_{j=1}^{p} D_j L^j,$$

L is the lag operator, and s and g are measured as deviations from their means. Letting $u_t' = (v_t, w_t)$, we have $E(u_t) = 0$ and

$$E(u_t u_{t-j}') = \begin{cases} \Omega & j = 0 \\ 0 & j \neq 0. \end{cases}$$

Under the additional assumption that u_t is distributed normally, the application of ordinary least squares to each equation in (3.17) produces asymptotically efficient estimates of the parameter vector

$$\theta' = [A_1, \ldots A_p, B_1, \ldots, B_p, C_1, \ldots, C_p, D_1 \ldots, D_p].$$

The resulting estimates, based on T observations, have the asymptotic distribution

$$\sqrt{T}(\hat{\theta} - \theta) \sim N[0, (\Omega \otimes M^{-1})]$$

where $M = \text{plim}(X'XT^{-1})$ and the matrix X contains the observations $(s_{t-1}, \ldots, s_{t-p}, g_{t-1}, \ldots, g_{t-p})$.

The derivation of the test statistics is facilitated by expressing (3.17) as a first order system in companion form as

$$
\begin{vmatrix}
s_t \\
s_{t-1} \\
\cdot \\
\cdot \\
s_{t-p+1} \\
g_t \\
g_{t-1} \\
\cdot \\
\cdot \\
g_{t-p+1}
\end{vmatrix}
=
\begin{vmatrix}
A_1 A_2 \cdots A_p & B_1 & B_2 \cdots B_p \\
1 \; 0 \\
0 \; 1 \; 0 \\
 & \cdot \\
0 \\
C_1 C_2 \cdots C_p & D_1 & D_2 \cdots D_p \\
0 & 0 \; 1 \; 0 & 0 \\
 & \cdot \\
 & & 0
\end{vmatrix}
\begin{vmatrix}
s_{t-1} \\
s_{t-2} \\
\cdot \\
\cdot \\
s_{t-p} \\
g_{t-1} \\
g_{t-2} \\
\cdot \\
\cdot \\
g_{t-p}
\end{vmatrix}
+
\begin{vmatrix}
v_t \\
0 \\
\cdot \\
\cdot \\
0 \\
w_t \\
0 \\
\cdot \\
\cdot \\
0
\end{vmatrix}
$$

which can be written as

$$Y_t = A Y_{t-1} + \eta_t. \tag{3.18}$$

Expressing the null hypothesis that $g_t = E_t(s_{t+4})$ requires isolating g_t and s_{t+4} in terms of (3.18). Let e_1' be a $1 \times 2p$ row vector composed entirely of zeroes except for unity in the $(p + 1)$ element. Then

$$g_t = e_1' A Y_{t-1} + w_t. \tag{3.19}$$

From (3.18) we can write

$$Y_{t+4} = A^5 Y_{t-1} + \sum_{j=0}^{4} A^j \eta_{t+4-j}. \tag{3.20}$$

To isolate s_{t+4}, let e_2' be a $1 \times 2p$ row vector composed entirely of zeroes except for unity in the first element. Then from (3.20)

$$s_{t+4} = e_2' A^5 Y_{t-1} + \sum_{j=0}^{4} e_2' A^j \eta_{t+4-j}. \tag{3.21}$$

Subtracting (3.19) from (3.21) gives

$$s_{t+4} - g_t = (e_2' A^5 - e_1' A) Y_{t-1} + \sum_{j=0}^{4} e_2' A^j \eta_{t+4-j} - w_t. \tag{3.22}$$

Baillie, Lippens and McMahon [12] and Hakkio [93] then project (3.22) onto the time $t - 1$ information set such that the error terms in (3.22) are eliminated:

$$E_{t-1}(s_{t+4}) - E_{t-1}(g_t) = (e_2'A^5 - e_1'A)Y_{t-1}. \qquad (3.23)$$

By applying the expectation operator based on time $t - 1$ information to both sides of (3.3) and using the law of iterated projections, one finds a testable null hypothesis:

$$e_2'A^5 - e_1'A = 0. \qquad (3.24)$$

Baillie, Lippens and McMahon [12] test the restrictions in (3.24) with a Wald test.[17]

The null hypothesis in (3.24) is a $2p$ vector of parameter constraints.

$$r(\theta)' = e_2'A^5 - e_1'A, \qquad (3.25)$$

on the unrestricted parameter vector θ. If we consider $r(\theta)$ to be a function of the true parameters, then we can expand (3.25) evaluated at the estimated unrestricted parameters in a linear Taylor series around the true parameters

$$r(\hat{\theta}) \simeq r(\theta) + D'(\hat{\theta} - \theta), \qquad (3.26)$$

where $D' \equiv \partial r(\theta)/\partial \theta$. Since $r(\theta) = 0$ under the null hypothesis

$$\sqrt{T}r(\hat{\theta}) \sim N[0, D'(\Omega \otimes M^{-1})D] \qquad (3.27)$$

from above. Therefore, an appropriate test statistic of the restrictions in (3.25) is

$$W = r(\hat{\theta})'\{D'[\hat{\Omega} \otimes X'X^{-1}]D\}^{-1}r(\hat{\theta}) \qquad (3.28)$$

where $\hat{\theta}$ is the vector of unrestricted estimates and $\hat{\Omega}$ is estimated from the residuals of (3.17). Computation of the derivative D is facilitated by using results from matrix differentiation due to Schmidt [191], that

$$\frac{\partial R(A^l)}{R(A)} = \sum_{j=0}^{l-1} A'^j \otimes A^{l-1-j} \qquad (3.29)$$

---

[17] Silvey [196] demonstrates that the Wald test is asymptotically equivalent to a likelihood ratio test. A nice feature of the Wald test is that it only requires estimation of the unrestricted model while maximum likelihood requires estimates of the parameters of the restricted model as well.

where R denotes the row stacking operator. In the particular case at hand,

$$D = \left[\frac{\sum_{j=0}^{4} (e_2' A^{\prime j} e_2) A^{4-j}}{\sum_{j=0}^{4} (e_1' A^{\prime j} e_2) A^{4-j} - I} \right] \tag{3.30}$$

where $e_2' A^{\prime j} e_2$ selects the $(1, 1)$ element and $e_1' A^{\prime j} e_2$ selects the $(p + 1, 1)$ element of $A^{\prime j}$.

Baillie, Lippens and McMahon [12] use data for six currencies in terms of their value against the US dollar. For the United Kingdom, West Germany, Italy, and France observations cover the period 1 June 1973 to 8 April 1980. For Canada and Switzerland the same quality of data are only available from 1 December 1977 to 15 May 1980. Their sample is a set of weekly observations from the New York foreign exchange market. For each week a one month forward rate is sampled on Tuesday and a spot rate on Thurdsay. This provides 362 observations for the larger sample and 128 for the smaller one.

Notice that their procedure generally matches the forward rate with the spot rate in the future that would be used to cover an open forward position. Riehl and Rodriguez [175] describe the institutional nature of delivery on spot and forward foreign exchange contracts. A purchase of spot foreign currency is for delivery two business days in the future on what is called the spot value date. To find the delivery date on a one month forward contract one first finds the appropriate spot value date in two business days then checks whether that numerical date in the next month is a business day in both countries. If it is, that day is the forward value date. If the corresponding numerical date in the future month is not a legitimate value date, forward settlement occurs on the next available business day without going out of the month. Instead of going into the next month, the contract is settled on the first business day before the numerical date corresponding to the spot value date of the previous month.

As an example, consider spot and forward contracts that were written on Tuesday, 5 March 1985. Spot contracts were delivered on Thursday, 7 March, and one month forward contracts were delivered on Monday, 8 April since 7 April was a Sunday. If one had to cover an open forward position with an April spot contract, one would have had to enter the spot market on Thursday, 4 April.

Hence, one can think of the Thursday spot rate on April 4th as the future spot rate that was being predicted by the forward contract signed on Tuesday, 5 March.

Empirical implementation of the procedure in Baillie, Lippens and McMahon [12] requires choice of the lag length p in the vector autoregression (3.17). They use procedures described in Akaike [5] and Hosking [109] to determine the appropriate length, and they find that the choice of p that minimizes the Akaike Information Criterion (AIC) of

$$-2 \ln \text{likelihood} + 2(\text{number of parameters})$$

is generally supported by the alternative specification tests against a $(p - 1)$ or a $(p + 1)$ process even though Sawa [190] notes that the AIC is generally biased toward acceptance of the larger model. They also state that their results for the Wald test of the unbiasedness hypothesis are insensitive to the value of p actually chosen.

Although Baillie, Lippens and McMahon [12] discuss their model in terms of the natural logarithms of the levels of spot and forward rates, in line with our previous discussion, actual estimation is done in first differences of the variables. To see that this is still a correct test first difference (3.3) to find

$$g_t - g_{t-1} = E_t(s_{t+4}) - E_{t-1}(s_{t+3}). \tag{3.31}$$

Projecting both sides of (3.31) onto the $t - 1$ information set implies that $g_t - g_{t-1} = E_{t-1}(s_{t+4} - s_{t+3})$ by the law of iterated expectations, hence the specification in (3.18) is appropriate. What this approach does not capture is the restriction on the time t information set that $g_t = E_t(s_{t+4})$. This is a different and stronger restriction than $E_{t-1}(g_t) = E_{t-1}(s_{t+4})$. Maximum likelihood estimation of the system in (3.13) and (3.14) employing the stronger restriction could lead to different inference, and this might be an important result if only mild evidence against the null hypothesis had been found, but Baillie, Lippens and McMahon [12] find very strong evidence against the null hypothesis. Some of their results are reported in Table II which reproduces their Table I.

In all cases the value of the Wald statistic is substantially greater than the value of $\chi^2(2p)$ at the 0.005 marginal level of significance. Hence, for each currency relative to the US dollar there is a very

TABLE II

Wald tests from Baillie, Lippens and McMahon [12]

Currency	Value of p that was finally chosen	Q_1	Q_2	$H(20)$	$H(54)$	Value of p that minimized the AIC	Likelihood ratio tests				Wald test statistic
							l_{-2}	l_{-1}	l_1	l_2	
UK pound	9	66.67[b]	52.15	46.36	178.07	9	21.72[a]	9.645[b]	5.26	7.91	191.98[a]
Deutsche mark	6	51.86	44.69	77.59[b]	207.63	6	25.63[a]	19.39[a]	1.28	9.05	249.16[a]
Italian lira	10	29.79	52.50	50.20	147.26	7	14.69[c]	9.57[b]	6.48	14.16	150.59[a]
French franc	6	40.58	47.63	78.01[b]	186.00	11	33.30[a]	16.90[a]	5.20	10.18	258.87[a]
Canadian dollar	3	15.74	17.17	68.86	—	3	34.86[a]	13.05[b]	1.79	10.11	52.33[a]
Swiss franc	3	7.28	18.43	58.70	—	3	25.64[a]	11.98[b]	5.15	11.70	69.94[a]

Notes: Q_1 and Q_2 are computed from $m = 54$ lags for the UK, West Germany, Italy, and France; and for $m = 20$ lags for Canada and Switzerland. The symbols l_{-2}, l_{-1}, l_1 and l_2 denote $-2 \ln$ of the likelihood ratio for the choice of lag p; l_{-2} is a test that the model is $AR(p-2)$ against $AR(p)$, l_{-1} tests $AR(p-1)$ against $AR(p)$, l_1 tests $AR(p)$ against $AR(p+1)$, and l_2 tests $AR(p)$ against $AR(p+2)$. Thus, under the null hypothesis that the lower value of p is appropriate l_{-2} and l_2 have asymptotic χ_8^2 distributions and l_{-1} and l_1 have asymptotic χ_4^2 distributions. The sample period is June 1973 to April 1980. The superscripts a, b, and c denote significance at the 1 percent, 5 percent, and 10 percent levels respectively.

strong rejection of the unbiasedness hypothesis. These results are quite consistent with those of Hakkio (1981) who uses a similar approach and data from the Netherlands, West Germany, Canada, Switzerland, and the UK for April 1973 to May 1977.

Levy and Nobay [134] criticize the first differencing approach of Baillie, Lippens and McMahon [12]. Levy and Nobay [134] view the logarithms of spot rates and forward rates as a bivariate time series process that is clearly nonstationary. Two alternatives that produce stationary processes are to first difference both series or to first difference the logarithm of the spot exchange rate to work with the rate of depreciation and to subtract the spot rate from the forward rate to work with the forward premium. Levy and Nobay [134] note that the former approach assumes two unit roots in the bivariate process while the latter implies only one unit root. Their analysis is consistent with the latter in which case the differencing approach may induce a unit root in the moving average representation through overdifferencing which leads to long autoregressive representations. There is no difference in inference, however, since the Wald statistics of Levy and Nobay [134] also indicate strong evidence against the unbiasedness hypothesis.

An alternative spectral analytic approach has recently been undertaken by Quah and Ito [173]. Their approach is to examine the restrictions, such as (3.25), in the frequency domain. They derive their restrictions from uncovered interest rate parity treating the logarithm of the exchange rate and the levels of the foreign and domestic interest rates as jointly covariance stationary time series. Whether this assumption, the type of parameterization chosen, or the small sample properties of their estimators explains their failure to reject the unbiasedness hypothesis is an open issue.

In concluding this section, mention must be made of Bailey, Baillie and McMahon [10] which surveys the alternative econometric approaches to testing unbiasedness. They note that rather than sampling the data and using ordinary least squares as Frenkel and others have done, one could use seemingly unrelated regressions as was proposed by Geweke and Feige [82] with the sampled data. Alternatively, one could use overlapping data as in Hansen and Hodrick [97], which will be discussed below, to increase power. A third alternative, where computationally feasible, is to use one of the maximum likelihood approaches discussed here. They note that rejection of the hypothesis appears to be related to the use of more powerful techniques, although specification of the equation appears to be important as well.

3.8. Problems with maximum likelihood estimation

Although the evidence in the previous section can be viewed as strong evidence against the unbiasedness hypothesis, it is based on an auxiliary assumption of conditional homoscedasticity. Hsieh [111] and Cumby and Obstfeld [37] argue that this assumption is invalid in these markets, and they propose and implement procedures that do not require this auxiliary assumption.

The problem of conditional heteroscedasticity arises because the assumption of rational expectations in (3.4) does not necessarily imply a constant conditional variance of $\varepsilon_{t+k,k}$, the forecast error due to new information that occurs between time t and time $t + k$. If, as Frenkel and Levich [74] argue, foreign exchange markets are characterized by tranquil and turbulent periods, or as Mussa [165] argues, these markets are subject to periods of quiescence and periods of turbulence, then it seems a priori likely that conditional heteroscedasticity may characterize the data. Of course, observing that ex post forward exchange rate forecast errors are small during some periods and large during other periods is not a test for conditional heteroscedasticity. Moreover, as Cumby and Obstfeld [37] note, it is possible to test for conditional heteroscedasticity prior to estimation since the null hypothesis that the unbiasedness hypothesis is true is particularly simple in this case. Then, since the hypothesis of conditional homoscedasticity is typically rejected by the data, one can apply the heteroscedastic consistent covariance matrixes for the coefficients that are derived in Hansen [96], Hsieh [110], and Cumby, Huizinga and Obstfeld [35]. Since these procedures can all be thought of as variants of Hansen's [96] Generalized Method of Moments (GMM), I discuss derivation of these covariance matrixes after discussion of GMM. An additional feature of GMM estimation is the ease with which one can use data sampled at any frequency that forecasts any horizon. Exact maximum likelihood may be intractable in such environments. The next section develops the GMM estimator.

3.9. Hansen's generalized method of moments

In the formulation and estimation of rational expectations models one invariably encounters orthogonality conditions such as (3.4). The model provides a set of restrictions that can be used to estimate the unknown parameters and to test the statistical adequacy of the economic theory. In this section I discuss Hansen's [96] GMM in general terms since several of the following sections employ particular versions of GMM estimation. Throughout this section I will use the unbiasedness hypothesis and the analysis of Hansen and Hodrick [97, 98] as examples of this style of research. I do not develop any of the formal arguments that are necessary to characterize the asymptotic distribution theory of the GMM estimator.[18] Instead, a general overview is provided.

Consider tests of the unbiasedness hypothesis (3.3) as a problem in GMM. The unbiasedness hypothesis implies that $s_{t+k} - g_{t,k}$ is orthogonal to all time t information. Consider the stationary specification

$$s_{t+k} - g_{t,k} = x_t'\delta_0 + \varepsilon_{t+k,k}. \qquad (3.32)$$

In this case we can think of the model as supplying us with a set of orthogonality conditions that may be used in estimation. For ease of presentation, define $y_{t+k} \equiv s_{t+k} - g_{t,k}$ and let x_t be an r-dimensional vector of time t information. Also define the function $h(y_{t+k}, x_t, \delta_0) = \varepsilon_{t+k,k}$, where δ_0 is the true parameter vector. Under the null hypothesis $\varepsilon_{t+k,k}$ is orthogonal to any variable in the time t information set. Hence, any vector of time t information x_t can be used to form a set of r orthogonality conditions that are defined by the function $f(y_{t+k}, x_t, \delta)$ as

$$f(y_{t+k}, x_t, \delta) \equiv h(y_{t+k}, x_t, \delta) \otimes x_t, \qquad (3.33)$$

where $E[f(y_{t+k}, x_t, \delta_0)] = 0$. Notice that under the null hypothesis $\varepsilon_{t+k,k}$ is serially correlated if $k > 1$. Since $\varepsilon_{t+k,k}$ is orthogonal to time t information, there are k autocovariances of the ε process including the zero order that are not equal to zero. Serial correlation complicates the estimation, but the consistency of the asymptotic covariance matrix of the estimator is not affected.

---

[18] While Hansen's [96] paper is the best reference for complete technical details, a good introduction to the estimation strategy in rational expectation environments is Cumby, Huizinga and Obstfeld [35].

Since the true parameter vector is unknown, a GMM estimator can be constructed by defining the function $g_0(\delta) \equiv$
$E[f(y_{t+k}, x_t, \delta)]$. This function has a zero at $\delta = \delta_0$, since it is equal
to zero under the null hypothesis. The method of moments
estimator of the function $g_0(\delta)$ for a sample of size T is

$$g_T(\delta) = T^{-1} \sum_{t=1}^{T} f(y_{t+k}, x_t, \delta). \tag{3.34}$$

The parameter vector, δ_T can be estimated by minimizing the
criterion function

$$J_T(\delta) = g_T(\delta)' W_T g_T(\delta), \tag{3.35}$$

where W_T is an r by r symmetric, positive definite weighting matrix
of the orthogonality conditions. In general the number of orthogonality conditions exceeds the number of parameters to be
estimated, otherwise δ_T could be chosen to set $g_T(\delta)$ equal to zero.
The choice of W_T defines the metric used in making g_T "close" to
zero.

When the orthogonality conditions in (3.33) are formed just with
the vector x_t, the GMM estimator of the parameters of (3.32) does
not depend on the choice of W_T, and the GMM estimator of δ_0
collapses to the ordinary least squares calculation.

Hansen [96] demonstrates that the asymptotic covariance matrix
of the parameters depends on the choice of the weighting matrix,
W_T, and he describes how W_T may be chosen "optimally" in
the sense of constructing an estimator of the parameters whose
asymptotic covariance matrix is the smallest among the class of
estimators that exploits that set of orthogonality restrictions. This is
an important result because the availability of data and computational tractability often restrict the choice of instrumental variables
and the number of orthogonality restrictions.

The derivations of the covariance matrix of the parameters and of
the optimal choice of weighting matrix require the specification of
some additional matrixes. Assume that the h function is
differentiable, which it is in this case, and define the matrix

$$D_0 \equiv E\left[\frac{\partial h}{\partial \delta}(y_{t+k}, z_t, \delta_0) \otimes x_t\right], \tag{3.36}$$

which will have full rank if x_t is chosen appropriately. Also, define

$$S_0 \equiv \sum_{j=-k+1}^{k-1} E[f(y_{t+k}, x_t, \delta_0)f(y_{t+k-j}, x_{t-j}, \delta_0)'] \qquad (3.37)$$

where the number of population autocovariances is determined by the order of the autocorrelation in the moving average term $\varepsilon_{t+k,k}$. Then Hansen's [96] Theorems 3.1 and 3.2 imply that the smallest asymptotic covariance matrix for an estimator δ_T that minimizes (3.35) can be chosen by letting $W_0^* = S_0^{-1}$, where W_T is assumed to converge almost surely to the constant limiting matrix W_0. The results in Hansen [96] can then be used to demonstrate that $\sqrt{T}(\delta_T - \delta_0)$ is normal with mean zero and an asymptotic covariance matrix for the parameters given by $(D_0'W_0^*D_0)^{-1}$.

In order to implement this procedure and to construct valid statistical inference, consistent estimators of D_0 and W_0^* are necessary. These are estimated by using their sample counterparts:

$$D_T = T^{-1} \sum_{t=1}^{T} \frac{\partial h}{\partial \delta}(y_{t+k}, x_t, \delta_T) \otimes x_t,$$

$$R_T(j) = T^{-1} \sum_{t=1+j}^{T} f(y_{t+k}, x_t, \delta_T)f(y_{t+k-j}, x_{t-j}, \delta_T)', \qquad (3.38)$$

$$W_T^* = \left\{ R_T(0) + \sum_{j=1}^{k-1} [R_T(j) + R_T(j)'] \right\}^{-1}.$$

Note that actual estimation of W_T^* requires a consistent estimator, δ_T, of δ_0. This can be obtained from an initial GMM estimation which does not impose the optimal weighting matrix. Once consistent estimates of δ_0 and the ε_t's are obtained, one can estimate δ_T^* which will have the smallest asymptotic covariance matrix for the class of estimators that exploit the same orthogonality conditions.[19]

---

[19] Two points are important to note. First, the construction of W_T in (3.38) is not guaranteed to be positive definite by construction. An alternative spectral estimator of W_T^* that does not have this potentially troublesome property is available (see Hansen [96]), but it does not impose the finite sum in defining S_0 in (3.37). Other alternatives are the estimators proposed by Cumby, Huizinga and Obstfeld [35] and Newey and West [166]. Second, although the estimator is optimal in the sense described in the text, this is a relatively narrow definition of optimality in the sense that different choices of instrumental variables may yield different inference. The general solution to the optimal choice of instruments may also be infeasible since it requires the researcher to specify the complete stochastic properties of the economic environment.

In linear environments such as (3.32), GMM estimators and their asymptotic standard errors are particularly easy to derive and to implement empirically. To see this, write the T observations on (3.32) as

$$y = X\delta_0 + \varepsilon \qquad (3.39)$$

where y is the T-dimensional vector of observations on y_{t+k}, X is the $T \times r$ matrix of observations on x_t and ε is the T-dimensional vector of error terms, $\varepsilon_{t+k,k}$. When $k > 1$, the error term is serially correlated up to lag $k - 1$, but the contemporaneous orthogonality between x_t and $\varepsilon_{t+k,k}$ and the assumed stationarity of the processes are sufficient to guarantee that ordinary least squares provides a consistent estimator of δ_0. Unfortunately, these parameter estimates are relatively inefficient compared to estimators that exploit more orthogonality conditions, and the asymptotic covariance matrix reported by standard OLS regression packages is inconsistent.

Hansen and Hodrick [97] also note that when the regressors in (3.39) are not strictly exogenous, as they are not in this particular application, use of generalized least squares leads to inconsistent estimators. Strict exogeneity requires $E(\varepsilon_{t+k,k} \mid x_t, x_{t-1}, x_{t+1}, \ldots) = 0$. Cumby, Huizinga and Obstfeld [35] provide a demonstration of why generalized least squares procedures leads to inconsistent parameter estimates. Before considering the construction of the GMM covariance matrix of the parameters, consider the following example that demonstrates why GLS is inconsistent.

Assume that $k = 2$ in (3.39), in which case $E(\varepsilon\varepsilon')$ has the same form as the covariance matrix of a first order moving average (MA) process since $E(\varepsilon_{t+k,k}\varepsilon_{t+k-j,k}) = 0$ for $j \geqslant 2$. Consequently, we can write

$$\varepsilon_t = v_t - \lambda v_{t-1} \qquad (3.40)$$

where v_t is a serially uncorrelated white noise process and $|\lambda| < 1$.[20]

---

[20] Since $E(\varepsilon_t^2) = (1 + \lambda^2)\sigma_v^2$ and $E(\varepsilon_t\varepsilon_{t-1}) = -\lambda\sigma_v^2$, λ is the real root of the equation $x^2 + [E(\varepsilon_t^2)/E(\varepsilon_t\varepsilon_{t-1})]x + 1 = 0$ that satisfies $|\lambda| < 1$. Since $E(\varepsilon_t^2)/E(\varepsilon_t\varepsilon_{t-1}) > 2$ by the Cauchy–Schwartz inequality, and because the product of the roots equals 1, both roots are real with one of them being λ and the other λ^{-1}. Hence, the requirement $|\lambda| < 1$ specifies the root that is smaller in absolute value.

Since $|\lambda| < 1$, generalized least squares can be thought of as the application of ordinary least squares to the filtered data

$$(1 - \lambda L)^{-1} y = (1 - \lambda L)^{-1} X \delta_0 + (1 - \lambda L)^{-1} \varepsilon. \qquad (3.41)$$

The regressors in (3.41) are a distributed lag of the regressors in (3.39) since $(1 - \lambda L)^{-1} x_t = \sum_{j=0}^{\infty} \lambda^j x_{t-j}$, while the error term in (3.41) is a distributed lag of current and past $\varepsilon_{t+k,k}$. Consequently, the new error term need not be uncorrelated with the filtered regressors making generalized least squares inconsistent. This occurs because it is generally the case that $E(x_t \varepsilon_{t+k-j,k}) \neq 0$ for $j > 1$ when x_t is not strictly exogenous.

When the vector of instruments is chosen just to be x_t, the sample estimator in (3.34) is

$$g_T(\delta) = X' \varepsilon / T, \qquad (3.42)$$

and the weighting matrix is a consistent estimate of the inverse of

$$\Omega = \lim_{T \to \infty} E(X' \varepsilon \varepsilon' X) / T. \qquad (3.43)$$

To see this equivalence, let $w_t' \equiv (\varepsilon_{t+k,k} x_t')$ be the realization of the function corresponding to the orthogonality condition (3.33). Then, under the regularity conditions stated in Hansen [96], w_t is a stationary, ergodic process. The value of $E(X' \varepsilon \varepsilon' X) / T$ for a sample of size T is found as

$$\frac{1}{T} E(X' \varepsilon \varepsilon' X) = \left(\frac{1}{T} \right) E \left\{ w_1 \sum_{t=1}^{T} w_t' + w_2 \sum_{t=1}^{T} w_t' + \cdots + w_T \sum_{t=1}^{T} w_t' \right\}$$

$$= \sum_{j=-k+1}^{k-1} \frac{T - |j|}{T} E(w_t w_{t-j}'), \qquad (3.44)$$

since $E_t(w_t w_{t-j}') = 0$ for $j > k$ and the law of iterated expectations applies. Taking the limit of (3.44) as $T \to \infty$ implies

$$\Omega = \sum_{j=-k+1}^{k-1} E(w_t w_{t-j}'). \qquad (3.45)$$

Then, recognize that $R_T(j)$ in (3.38) is the sample estimator of $E(w_t w_{t-j}')$ and that $R_T(-j) = R_T(j)'$. These facts imply that W_T^* in (3.38) is a sample estimate of Ω^{-1}.

The criterion function in (3.35) can therefore be written as

$$J_T(\delta) = (X'y - X'X\delta)' W_T^* (X'y - X'X\delta) \qquad (3.46)$$

and minimization of (3.46) with respect to δ gives

$$\delta_T = (X'X)^{-1}X'y, \tag{3.47}$$

the ordinary least squares estimator of δ_0.

The asymptotic covariance matrix is also easy to calculate. From (3.38) recognize that $D_T = X'X/T$ when the instruments are chosen just to be the regressors. Consequently, the asymptotic covariance matrix $(D_0'W_0^*D_0)^{-1}$ is estimated as

$$(X'X/T)^{-1}W_T^*(X'X/T)^{-1}. \tag{3.48}$$

It should be recognized that the estimator in (3.48) does not impose an assumption of conditional homoscedasticity. This assumption is imposed by Hansen and Hodrick [97], and it can be expressed as

$$E(\varepsilon_{t+k,k}\varepsilon_{t+k-j,k} \mid x_t, \varepsilon_{t,k}, x_{t-1}, \varepsilon_{t-1,k}, \ldots) = R_\varepsilon(j) \tag{3.49}$$

for $0 \leqslant j < k$, where the $R_\varepsilon(j)$ is a constant. If we also define the autocovariances $E(x_t x_{t+j}' \mid x_t, \varepsilon_{t,k}, x_{t-1}, \varepsilon_{t-1,k}, \ldots) = R_x(j)$, then S_0 in (3.37) under the assumption of homoscedasticity is

$$S_0 = \sum_{j=-k+1}^{k-1} R_\varepsilon(j) \otimes R_x(j) \tag{3.50}$$

which can be consistently estimated by using the facts that $R_\varepsilon(j) = R_\varepsilon(-j)$ and $R_x(j) = R_x(-j)'$ and constructing the moment estimators of $R_x(j)$ and $R_\varepsilon(j)$ given by

$$R_x^T(j) = \left(\frac{1}{T}\right) \sum_{t=j+1}^{T} x_t x_{t-j}'$$
$$R_\varepsilon^T(j) = \left(\frac{1}{T}\right) \sum_{t=j+1}^{T} \hat{\varepsilon}_{t+k,k}\hat{\varepsilon}_{t+k-j,k}. \tag{3.51}$$

3.10. GMM results with weekly data and three-month forecasts

Hansen and Hodrick [97] utilize the modified ordinary least squares approach discussed above in their study. Their specification of the unbiasedness hypothesis is also in natural logarithms, and they use weekly data with three month forward rates. As a consequence, their forecast interval k is thirteen. Hansen and Hodrick [97] use two specifications in their tests. One corresponds to what is commonly called a weak form efficiency test because it only uses the past history of the variable being forecast as right-hand-side variables. This first specification is

$$s^i_{t+13} - g^i_{t,13} = \delta_{i0} + \delta_{i1}(s^i_t - g_{t-13,13})$$
$$+ \delta_{i2}(s^i_{t-1} - g^i_{t-14,13}) + \varepsilon^i_{t+13,13} \qquad (3.52)$$

where the index i covered the currencies of Canada, West Germany, France, the United Kingdom, Switzerland, Japan, and Italy relative to the US dollar, and where the null hypothesis that the forward rate is an unbiased predictor of the future spot rate is $\delta_{i0} = \delta_{i1} = \delta_{i2} = 0$. The evidence in this case is mixed. There is only strong evidence against the null hypothesis for the Deutsche mark-US dollar exchange rate for which one can reject the null hypothesis at all marginal levels of significance larger than 0.008.

Additional evidence against the unbiasedness hypothesis is available, though, in what is commonly called a semi-strong form efficiency test because it utilizes other publicly available data as right hand side variables. The specification in this case is

$$s^i_{t+13} - g^i_{t,13} = \delta_{i0} + \sum_{j=1}^{5} \delta_{ij}(s^j_t - g^j_{t-13,13}) + \varepsilon^i_{t+13,13} \qquad (3.53)$$

where each of the forecast errors for the US dollar prices of the Canadian dollar, the Deutsche mark, the French franc, the UK pound, and the Swiss franc were regressed on the currently occurring forecast errors from the five currencies. This approach yields evidence against the unbiasedness hypothesis for the Canadian dollar, the Swiss franc, and the Deutsche mark. Hansen and Hodrick [97] note that the three month forecast errors produce a relatively large overlap of data. While using all of the available data increases the degrees of freedom compared to the alternative of sampling the data to yield a nonoverlapping sample, they note the desirability of using either one month or other shorter forecast intervals in order to increase the degrees of freedom from nonoverlapping data.

Hansen and Hodrick [97] also provide a proof that using the overlapping data and the GMM approach described above with the asymptotic covariance matrix given by (3.48) is superior to sampling the data to produce a nonoverlapping data set to which ordinary least squares and the traditional computation of the asymptotic covariance matrix could be applied directly.

3.11. Results with weekly sampling and a one-week forecast interval

In order to improve the degrees of freedom associated with the tests of unbiasedness, Cumby and Obstfeld [36] and Hsieh [111] use weekly data on Eurocurrency interest rates to construct an implicit one week forward rate. From (2.11) we know that

$$G_t^{ij} = S_t^{ij} R_t^j / R_t^i \qquad (3.54)$$

where S_t^{ij} and G_t^{ij} are the spot and forward exchange rates of currency j per unit of currency i, R_t^i is one plus the known nominal one-period interest rate in currency i, and R_t^j is one plus the known nominal one-period interest rate in currency j. Because interest rate parity holds quite well up to transactions costs that are quite small, as McCormick [152] demonstrates, the effective forward rates defined by the right-hand side of (3.41) can be appropriately used in tests of this kind.

Hsieh [111] compares the inference that arises when one uses the asymptotic covariance matrix that imposes an assumption of conditional homoscedasticity versus one that does not impose this assumption as was discussed above in the deviation of (3.48). With nonoverlapping data Hsieh's specification is

$$y_{t+1} = x_t' \delta_0 + \varepsilon_{t+1} \qquad (3.55)$$

and where $y_{t+1} \equiv s_{t+1} - g_t$ is the logarithmic forward rate forecast error and the variables in x_t vary with alternative specifications. Under the null hypothesis ε_{t+1} is serially uncorrelated and $\sqrt{T}(\delta_T - \delta_0)$ is asymptotically normal with mean zero and covariance matrix estimated by

$$V_{HC} = (X'X)^{-1} \left(\sum_{t=1}^{T} \hat{\varepsilon}_{t+1}^2 x_t x_t' \right) (X'X)^{-1} T, \qquad (3.56)$$

when conditional homoscedasticity is not assumed, and by

$$V_{OC} = \hat{\sigma}^2 (X'X)^{-1} T, \qquad (3.57)$$

when conditional homoscedasticity is assumed. As in the previous section, X is the $T \times r$ matrix of T observations on the r-dimensional vector x_t. In (3.57) $\hat{\sigma}^2$ is the sum of squared residuals, $\hat{\varepsilon}_{t+1}^2$, from the ordinary least squares regression (3.55) divided by the sample size. No adjustment need be made for degrees of freedom since V_{OC} is only justified asymptotically.

Hsieh [111] notes that

$$V_{HC} - V_{OC} = (X'X)^{-1} \left[\sum_{t=1}^{T} (\hat{\varepsilon}_{t+1}^2 - \hat{\sigma}^2) x_t x_t' \right] (X'X)^{-1} T. \quad (3.58)$$

The expression (3.58) goes to zero asymptotically when ε_{t+1}^2 and $x_t x_t'$ are not correlated. If they are positively correlated, V_{OC} tends to underestimate the true covariance matrix of the parameters, while if they are negatively correlated, V_{OC} tends to overestimate the former covariance matrix. Since V_{HC} is a consistent estimator whether there is heteroscedasticity or not, Hsieh argues forcefully that use of V_{HC} is the appropriate way to conduct statistical inference. It is also quite striking how differet inference is with the two matrixes.

Consider Table III which reproduces results of Hsieh's [111] Table 4. The specification is

$$s_{t+1}^i - g_t^i = \delta_0 + \sum_{j=1}^{8} \delta_{ij}(s_t^j - s_{t-1}^j) + \sum_{j=1}^{8} \beta_{ij}(g_t^j - s_t^j) + \varepsilon_{t+1}^i. \quad (3.59)$$

The eight exchange rates of the study are the US dollar values of the currencies of France, Germany, the United Kingdom, Switzerland, the Netherlands, Canada, Italy, and Japan.

The chi-square statistics in Table III provide some very interesting results. When the covariance matrix that imposes the assumption of conditional homoscedasticity is used to construct the test statistics, only somewhat weak evidence is found against the unbiasedness hypothesis. Only four currencies have test statistics that are larger than the value associated with the ten percent marginal level of significance.[21] A very different story emerges when

---

[21] Of course, as Geweke and Feige [82] note, a joint test of the foreign exchange market rather than eight separate currency markets requires a joint test that all 136 coefficients in Table III are equal to zero. Such a test would be likely to provide strong evidence against unbiasedness. Intuitively, the reason behind this statement is that finding four test statistics out of eight that have ten percent significance is a highly improbable event if all of the coefficients are truly zero.

R. J. HODRICK

TABLE III
Heteroscedastic consistent tests from Hsieh [111]

$$s_{t+1}^i - g_t^i = \delta_0 + \sum_{j=1}^{8} \delta_{ij}(s_t^j - s_{t-1}^j) + \sum_{j=1}^{8} \beta_{ij}(g_t^j - s_t^j) + \varepsilon_{t+1}^i$$

| | Tests of $\delta_0 = \delta_{ij} = \beta_{ij} = 0 \; \forall j$ | |
Currency	OC	HC
French franc	27.48*	41.20***
Deutsche mark	28.15*	40.00***
UK pound	17.68	35.58***
Swiss franc	20.45	37.46***
Dutch guilder	27.54	35.42***
Canadian dollar	24.44	31.91***
Italian lira	24.34	33.68***
Japanese yen	24.86*	28.57**

Period: 29 September 1978 to 24 April 1981 (139 weekly observations)

Note: OC and HC are chi-square statistics with 17 degrees of freedom for tests of the null hypothesis that all coefficients are equal to zero. They are constructed from V_{OC} and V_{HC}, respectively. Critical values are 24.7690, 27.5871, and 33.4087 at the 10, 5, and 1 percent marginal levels of significance, respectively. Rejections of the null hypothesis at these levels of significance are indicated by one, two, or three *'s, respectively.

test statistics are constructed from the covariance matrix that does not impose the additional auxiliary assumption of conditional homoscedasticity. Seven of the currencies have test statistics that are larger than the value associated with the one percent marginal level of significance, and the one for the Japanese yen is larger than the value for the five percent level. Since the heteroscedastic covariance matrix is a consistent estimator of the true covariance matrix, the appropriate inference to draw is that there is strong evidence against the unbiasedness hypothesis.

If we take V_{HC} as the correct covariance matrix, V_{OC} is apparently biased in this case. Because the chi-square test statistics are constructed from the inverses of V_{OC} or V_{HC} as $(\delta_T' V^{-1} \delta_T)T$, it must be the case that V_{OC} overestimates the true covariance matrix. From (3.58) this occurs when the squared regressors are negatively correlated with the squared forecast error. Further investigation of this proposition would seem to be a worthwhile enterprise, since modeling the conditional variance may be a fruitful direction to pursue in order to understand the nature of the rejection of the unbiasedness hypothesis and to determine whether the rejection is due to a time varying risk premium.[22]

3.12. Testing for conditional homoscedasticity

Cumby and Obstfeld [37] stress the ease with which one can test for conditional homoscedasticity in these simple environments, and they advocate use of the conditional heteroscedastic covariance matrix in conducting inference.

Cumby and Obstfeld [37] consider estimation of

$$s_{t+k} - s_t = \alpha + \beta(g_{t,k} - s_t) + \varepsilon_{t+k,k} \tag{3.60}$$

with more instrumental variables than just a constant and the right-hand-side variable. Prior to the estimation of α and β, they test for conditional homoscedasticity by imposing the null hypothesis that $\alpha = 0$ and $\beta = 1$. In this case $\varepsilon_{t+k,k} = s_{t+k} - g_{t,k}$ is observable, and the assumption that $E_t(\varepsilon^2_{t+k,k}) = \sigma^2_\varepsilon$ can be tested in the specification proposed by White (1980),

$$\varepsilon^2_{t+k,k} = \delta_0 + \delta_1(g_{t,k} - s_t) + \delta_2(g_{t,k} - s_t)^2 + v_{t+k,k}. \tag{3.61}$$

Under the hypothesis of conditional homoscedasticity, the expected value of the squared forward rate forecast error at time t is constant. Therefore, the test is $\delta_1 = \delta_2 = 0$ in (3.61). Any variable in the information set is a legitimate instrumental variable that can be used in the test, and Cumby and Obstfeld [37] chose as instruments the time t forward premiums and the squared values of the forward premiums for the five US dollar values of the currencies in their study.

Cumby and Obstfeld [37] estimate (3.60) and (3.61) with what Cumby, Huizinga and Obstfeld [35] refer to as two-step, two-stage,

---

[22] In Section 5 the model of Domowitz and Hakkio [42] is discussed. They relate the conditional variance of the forecast error directly to a model of the risk premium and impose a particular structure on the conditional heteroscedasticity in their estimation. McCurdy and Morgan [154] also model the heteroscedasticity in their study of futures prices which is discussed in Section 7.

least squares (2S2SLS). They also note the equivalence of 2S2SLS with GMM. In terms of the derivation of the parameter estimators of (3.32) think of having an r-dimensional vector of instruments denoted z_t with the property that they are contemporaneously orthogonal to the error terms in (3.60) or (3.61). For ease of presentation let either of these equations be represented by

$$y_t = x_t'\delta_0 + \varepsilon_t, \tag{3.62}$$

and the orthogonality property is $E(\varepsilon_t z_t) = 0$.

A GMM or 2S2SLS estimator of δ_0 based on the r orthogonality conditions minimizes the sum of squared residuals of

$$R^{-1}Z'y = R^{-1}Z'X\delta_0 + R^{-1}Z'\varepsilon \tag{3.63}$$

where $RR' = \Omega_T$, which is an estimator of

$$\Omega = \lim_{T\to\infty} E(Z'\varepsilon\varepsilon'Z)/T.$$

The estimate of δ_0 is

$$\delta_T = (X'Z\Omega_T^{-1}Z'X)^{-1}X'Z\Omega_T^{-1}Z'y \tag{3.64}$$

and $\sqrt{T}(\delta_T - \delta_0)$ converges in distribution to a normal random variable with mean zero and variance covariance matrix that is consistently estimated by

$$[(X'Z/T)\Omega_T^{-1}(Z'X/T)]^{-1}, \tag{3.65}$$

with Ω_T estimated from the sample moments corresponding to $E(z_t\varepsilon_t\varepsilon_{t-j}z_{t-j}')$. As in the derivation of (3.45) the number of moments included in Ω_T depends on the null hypothesis. In (3.60) and (3.61) the residuals are autocorrelated up to lag k due to the overlapping nature of the forecasts.

The results of the Cumby and Obstfeld [37] tests for heteroscedasticity are presented in Table IV which is taken from their Table 3.3. Notice the exceedingly strong evidence against conditional homoscedasticity since the probability of finding a value of a $\chi^2(2)$ greater than 10.5966 is 0.005. Thus, the hypothesis of conditional homoscedasticity is rejected at very small marginal levels of significance for the exchange rates of the US dollar versus the UK pound, the Deutsche mark, the Swiss franc, and the Japanese yen. Somewhat surprisingly given the strength of the evidence for the

TABLE IV
Tests of conditional homoscedasticity in
forward-rate forecast errors from Cumby and
Obstfeld [37]

Currency	Test statistic
UK pound	308.13
Deutsche mark	26.38
Swiss franc	13.20
Canadian dollar	2.57
Japanese yen	141.05

Note: The test statistic is distributed as a
chi-square with two degrees of freedom. The
specification is given by (3.61). Critical
values are 5.99147 for the 0.05 level and
9.21034 for the 0.01 level.

other currencies, there is little evidence against conditional homoscedasticity in the case of the Canadian dollar.

The estimation of (3.60) is conducted with a set of instruments
that consists of the five forward premiums for the currencies in the
study. Weekly data for the time period January 1976 to June 1981
are employed, and the forecast interval is three months or thirteen
weeks. The results are presented in Table V.

TABLE V
Tests of unbiasedness from Cumby and Obstfeld [37]

Currency	$\hat{\alpha}$	$\hat{\beta}$	Test statistic
UK pound	0.0086 (0.0156)	−0.2881 (0.9741)	16.16
Deutsche mark	0.0214 (0.0113)	−0.7815 (1.1579)	3.59
Swiss franc	0.481 (0.0214)	−2.2145 (1.1177)	9.11
Canadian dollar	−0.0076 (0.0023)	0.8285 (0.7922)	12.44
Japanese yen	0.0311 (0.0097)	−2.8316 (0.6740)	41.58

Note: Standard errors appear in parenthesis. The test statistic is
distributed as a chi-square with two degrees of freedom. The specification
is given by (3.60). Critical values are 5.99147 for the 0.05 level and 9.21034
for the 0.01 level.

Notice that in each case, except the Canadian dollar, the estimate of β in (3.60) is negative although the standard errors are relatively large compared to the coefficients. This is in strong contrast to the prediction of the unbiasedness hypothesis that $\beta = 1$. Negative β's in specifications like (3.60) are also found in studies using monthly data by Bilson [14] and Fama [59]. These works are discussed in detail below. Here I merely note the consistency of the rejection of the unbiasedness hypothesis by Cumby and Obstfeld [37] with other research. For the US dollar value of the UK pound, the Canadian dollar, and the Japanese yen, there is sufficiently strong evidence to reject the implications of the unbiasedness hypothesis at the one percent marginal level of significance, and for the Swiss franc the value of the test statistic is only slightly below the critical value of the one percent level. Cumby and Obstfeld [37] note that the rejection for the Canadian dollar appears to be due almost entirely to the constant term which is negative and almost three times its standard error. Although the results for the Deutsche mark lend support to the unbiasedness hypothesis, Cumby and Obstfeld [37] note that the rejections for this currency have been particularly strong in other studies as was documented above.

4. ALTERNATIVE INTERPRETATIONS OF REJECTIONS OF THE UNBIASEDNESS HYPOTHESIS

In Section 3 considerable evidence for a variety of currencies and sample periods is presented that indicates a strong rejection of the proposition that the forward exchange rate is an unbiased predictor of the future spot rate. In this section an examination of two alternative interpretations of this finding is discussed. First, the interesting variance decomposition suggested by Fama [59] is analyzed. Then, the profitability of various trading strategies is examined, and some contentions that the market is inefficient are addressed.

There are basically three ways that one can interpret the evidence from the preceding section. These interpretations are not mutually exclusive because some combination of the three could also be an explanation. These three propositions correspond to well-defined positions within the profession on this issue as is noted by Hodrick and Srivastava [106].

The first position is occupied by those who continue to support the unbiasedness hypothesis by arguing along one of two lines. Either there is a statistical problem with the data that makes the application of asymptotic distribution theory inappropriate and the analysis subject to severe small sample bias, or it is argued that the unbiasedness hypothesis cannot be rejected until we have an alternative model of a time varying risk premium that is not rejected by the data. This latter position is related to Frankel's [67] argument that rejection of the unbiasedness hypothesis in studies like those of the previous section fails to provide evidence about the nature of the time varying risk premium that separates forward exchange rates from expected future spot rates because the alternative model of risk averse behavior is not specified.

There is definite validity in both sets of arguments. Geweke and Feige [82] argue that the unbiasedness hypothesis is probably not literally true and that statistical tests of the theory ought to be powerful enough to reject the null hypothesis while simultaneously providing an indication of why the hypothesis is not true. A discussion of their analysis of the rejection of unbiasedness is examined below when Bilson's [14] contention that the market is inefficient is discussed. Other early studies such as Hansen and Hodrick [97] also are careful to note that the rejection of unbiasedness has not necessarily isolated a time invariant forecasting rule. Hodrick and Srivastava [106] address this issue in their analysis of Bilson's [14] trading rule, and this is discussed below as well.

The second position within the profession is occupied by those who view the rejection of the unbiasedness hypothesis as evidence against the efficiency of the market. Bilson [14] investigates the 'speculative efficiency' hypothesis which is the unbiasedness proposition. He finds that regression equations like those from Section 3 can be used to develop a trading strategy with the property that [14, p. 449] "the profit/risk ratio appears to be too large to be accounted for in terms of risk aversity." Earlier, McKinnon [156, p. 156] argued

that the supply of private capital for taking net positions in either the forward or spot markets is currently inadequate. Exchange rates can move sharply in response to random variations in the day-to-day demands of merchants or from monetary disturbances. Once a rate starts to move because of some temporary perturbation, no prospective speculator is willing to hold an open position for a significant time interval in order to bet on a reversal—whence the large daily and monthly movements in the foreign exchanges and sometimes high bid-ask spreads. Bandwagon psychologies result from the general unwillingness of participants to take net positions against near-term market movements that are necessarily accentuated by the behavior of nonspeculative merchants.

This line of reasoning is taken seriously by Dooley and Shafer [44, 45] who investigate the profitability of filter rules that borrow depreciating currencies and lend appreciating currencies. Their analysis is described below. Based on the out-of-sample profitability of certain filters, Dooley and Shafer [45, p. 68] conclude, 'many currencies either were not efficient in their use of price information or real interest differentials were large and variable during the sample period'. Similarly, Sweeney [207, p. 178] argues that 'major exchange markets showed grave signs of inefficiency over the first 1830 days of generalized floating.' Clearly, this second position is taken seriously within the profession. These filter rule studies are addressed below.

The third position with respect to the rejection of the unbiasedness hypothesis is occupied by those who are attempting to develop and test tractable empirical models of risk premiums. These papers are discussed in Section 5. As noted above, these three positions are not mutually exclusive since problems with the small sample properties of the data hamper interpretation of all tests of market efficiency.

One way to interpret the rejection of the unbiasedness hypothesis that retains consistency with market efficiency is to argue that the evidence is consistent with time varying risk premiums. Fama [59] provides an interesting insight in this regard.

4.1. Fama's decomposition argument

In order to investigate the variability of risk premiums and expected rates of currency depreciation as well as their covariability, Fama [59] proposes a simple model for these measurements.

Conditional on market efficiency and rational expectations, Fama [59] argues that the forward exchange rate is equal to the expected future spot rate plus a risk premium as is demonstrated in the derivation of (2.16). Based on the analysis in Fama and Farber [60], and on the theoretical model of Stulz [202], Fama [59] specified his model in natural logarithms,

$$g_{t,k} = E_t(s_{t+k}) + p_{t,k} \tag{4.1}$$

where $p_{t,k}$ is the logarithmic risk premium. Fama [59] treats (4.1) as a definition of the risk premium, and in order to conduct statistical inference, he subtracts s_t from both sides of (4.1) to yield

$$g_{t,k} - s_t = E_t(s_{t+k} - s_t) + p_{t,k}. \tag{4.2}$$

The left-hand side of (4.2) is the forward premium, and the right-hand side is the expected rate of depreciation of the home currency relative to the foreign currency, where the exchange rates are home currency per unit of foreign currency, plus the risk premium that can be thought of as a percentage return to selling the foreign currency forward. In the analysis of this issue in Hodrick and Srivastava [108], the specification corresponding to (4.1) is in levels,

$$G_{t,k} = E_t(S_{t+k}) + P_{t,k} \tag{4.3}$$

and the statistical analysis is conducted by subtracting S_t from both sides of (4.3) and by dividing through by S_t,

$$(G_{t,k} - S_t)/S_t = E_t[(S_{t+k} - S_t)/S_t] + P_{t,k}/S_t. \tag{4.4}$$

Since there is extremely high correlation between the two definitions of the forward premium and the two definitions of the rate of depreciation, it is unlikely that any statistically significant, or more importantly, any economically significant difference arises from using (4.4) versus (4.3).[23]

Fama [59] reexamines regressions of the actual rate of depreciation on the forward premium that had been used to test the unbiasedness hypothesis, such as (3.47), in light of the specification of the forward premium in (4.2). His analysis considers the two

---

[23] Hansen and Hodrick [98] examine the correlation between $(S_{t+k} - G_{t,k})/S_t$ and $\ln(S_{t+k}) - \ln(G_{t,k})$ for the US dollar values of the French franc, the Japanese yen, the Swiss franc, the UK pound, and the Deutsche mark. They find the correlation between the two representations to be 0.999 for all five exchange rates for the sample period from February 1976 to December 1980. Hence, it is unlikely that differences in inference arise from using one specification rather than the other.

complementary regressions with nonoverlapping data given by

$$
\begin{aligned}
g_t - s_{t+1} &= \alpha_1 + \beta_1(g_t - s_t) + \varepsilon_{t+1}^1 \\
s_{t+1} - s_t &= \alpha_2 + \beta_2(g_t - s_t) + \varepsilon_{t+1}^2
\end{aligned}
\tag{4.5}
$$

which are clearly dependent since the stochastic regressor is the same in both equations and the sum of the dependent variables is the stochastic regressor. The complete complementarity of the regressions in (4.5) implies that $\hat{\alpha}_1 = -\hat{\alpha}_2$, that $\hat{\beta}_1 = 1 - \hat{\beta}_2$, and that $\hat{\varepsilon}_{t+1}^1 = -\hat{\varepsilon}_{t+1}^2$. Fama [59] argues that the usefulness of reporting both presentations is in interpreting the data. The data are one month forward rates with no overlap as is reflected in (4.5).

If the equations in (4.5) are viewed as linear predictors or projections, we know that ordinary least squares will isolate ε_{t+1}^1 and ε_{t+1}^2 as the components of $(g_t - s_{t+1})$ and $(s_{t+1} - s_t)$ that are orthogonal to the forward premium. Under appropriate regularity conditions, the probability limits of $\hat{\beta}_1$ and $\hat{\beta}_2$ are given by

$$
\begin{aligned}
\beta_1 &= C(g_t - s_{t+1}; g_t - s_t)/V(g_t - s_t) \\
\beta_2 &= C(s_{t+1} - s_t; g_t - s_t)/V(g_t - s_t)
\end{aligned}
\tag{4.6}
$$

where $C(\cdot;\cdot)$ and $V(\cdot)$ denote the unconditional covariance and variance, respectively. The assumption of rational expectations implies that $s_{t+1} - s_t = E_t(s_{t+1} - s_t) + v_{t+1}$ where v_{t+1} is orthogonal to all time t information as in (3.4). Hence, combining the rational expectations assumption with the decomposition of the forward premium in (4.2) implies that

$$
\begin{aligned}
C(g_t - s_{t+1}; g_t - s_t) &= C[p_t; E_t(s_{t+1} - s_t) + p_t] \\
&= C[p_t; E_t(s_{t+1} - s_t)] + V(p_t), \tag{4.7}
\end{aligned}
$$

$$
\begin{aligned}
C(s_{t+1} - s_t; g_t - s_t) &= C[E_t(s_{t+1} - s_t); E_t(s_{t+1} - s_t) + p_t] \\
&= C[p_t; E_t(s_{t+1} - s_t)] + V[E_t(s_{t+1} - s_t)] \tag{4.8}
\end{aligned}
$$

and

$$
\begin{aligned}
V(g_t - s_t) &= V[E_t(s_{t+1} - s_t)] + V(p_t) \\
&\quad + 2C[p_t; E_t(s_{t+1} - s_t)]. \tag{4.9}
\end{aligned}
$$

Hence,

$$
\beta_1 = \frac{C[p_t; E_t(s_{t+1} - s_t)] + V(p_t)}{V[E_t(s_{t+1} - s_t)] + V(p_t) + 2C[p_t; E_t(s_{t+1} - s_t)]}
\tag{4.10}
$$

and

$$\beta_2 = \frac{C[p_t; E_t(s_{t+1} - s_t)] + V(E_t(s_{t+1} - s_t))}{V[E_t(s_{t+1} - s_t)] + V(p_t) + 2C[p_t; E_t(s_{t+1} - s_t)]}. \quad (4.11)$$

The coefficients β_1 and β_2 describe roughly the degree of variability of the components of the forward premium. Fama [59] notes that only if the risk premium and the expected rate of depreciation are uncorrelated would β_1 be equal to the proportion of the variance of the forward premium due to variance of the risk premium, and under this condition β_2 would be equal to the proportion of the variance of the forward premium due to variance of the expected rate of depreciation. Since it is unlikely that the two components of the forward premium are uncorrelated, the covariance terms in (4.10) and (4.11) must be taken into account. Hence, information about the relative variability of the two components is masked somewhat by their covariability.

Notice, though, how the interpretation is affected if $\beta_2 < 0$. Because the denominator and the variance term in the numerator of (4.11) must be positive, a finding of $\beta_2 < 0$ implies that the covariance between the expected rate of depreciation and the risk premium must be negative and larger in absolute value than the variance of the expected rate of depreciation. Since the variance of the forward premium must be positive, this finding carries with it the additional implication that

$$V(p_t) > |C[p_t; E_t(s_{t+1} - s_t)]| > V[E_t(s_{t+1} - s_t)]. \quad (4.12)$$

It is in the sense of (4.12) that one can then conclude that the variance of the risk premium is greater than the variance of the expected rate of depreciation.

Fama [59] uses monthly data taken from the Harris Bank Data Base supported by the Center for Studies in International Finance at the University of Chicago. Spot exchange rates and thirty-day forward rates are sampled from Friday closes at four-week intervals for the US dollar values of nine major currencies. The currencies included in the study are from Belgium, Canada, France, Italy, Japan, the Netherlands, Switzerland, the United Kingdom, and

West Germany. The time period was from 31 August 1973 to 10 December 1982, or 122 nonoverlapping observations.[24]

Fama [59] estimates (4.5) with OLS and seemingly unrelated regression (SUR). The validity of his standard errors requires the additional assumptions of conditional homoscedasticity and no serial correlation. Both of these hypotheses are questionable in this situation, perhaps the former more than the latter given the evidence of Section 3.

Hodrick and Srivastava [108] discuss a potential bias in Fama's [59] analysis due to the nature of the error term in (4.5). Consider the composition of the second error term, ε_{t+1}^2, in (4.5). Since OLS provides the projection of $s_{t+1} - s_t$ onto a constant and $g_t - s_t$,

$$\varepsilon_{t+1}^2 = v_{t+1} + \mu_t \qquad (4.13)$$

where v_{t+1} is the rational expectations error, which is not serially correlated in a nonoverlapping sample, and μ_t is the error induced by the fact that the forward premium, in the presence of a risk premium, is not the time t conditional expectation of the rate of change of the exchange rate. Hodrick and Srivastava [108] argue that prior evidence indicates that μ_t is probably serially correlated. Their argument rests on the fact that in analyses of unbiasedness, such as those in Section 3 or their previous analysis (Hodrick and Srivastava [106]), one finds statistically significant explanatory power for the set of forward premiums for other currencies, which typically are highly serially correlated, in a regression equation like (4.5). Thus, it seems that the residual that is decomposed in (4.13) may be characterized by weak serial correlation.

---

[24] Note that the Harris Bank data cannot be aligned correctly because only Friday observations are available. Following the settlement rules described in Section 2, a Friday forward rate generally predicts the spot rate on a Tuesday thirty days in the future. This is because spot contracts entered into on Friday are settled the following Tuesday unless Monday is not a business day, and Friday forward contracts are settled on the date in the next month corresponding to the Tuesday spot value date if the date in the next month is a spot value day. This date in the next month is a Thursday if the current month has thirty days and a Friday if the current month has thirty-one days. The corresponding future spot rate that the forward rate predicts is therefore a Tuesday or a Wednesday two business days before the delivery day. Given the results of various studies that have aligned the data correctly, it seems unlikely that the measurement error induced by use of the Harris Bank data is a serious problem although better data can be purchased from Data Resources, Inc. who have daily data on spot and forward rates of various maturities since the beginning of the flexible rate period.

Since Fama [59] checks his residuals with standard time series tests for serial correlation and fails to detect evidence against the hypothesis of no serial correlation, it must be the case that the residual autocorrelation is too weak to emerge in the data with standard tests. Hodrick and Srivastava [108] therefore ask two questions. Why do the residuals of (4.5) satisfy standard tests for serial correlation, and given that statistically significant serial correlation cannot be detected, is there the possibility of significant bias in the estimation of the standard errors?

To answer the first question it is important to realize that the standard test for serial correlation compares an estimated autocorrelation coefficient to its asymptotic standard error under the null hypothesis of no serial correlation. The standard error of the estimated autocorrelation coefficients is $1/\sqrt{T}$ where T is the sample size. With 122 monthly observations, the standard error is approximately 0.09 implying that an autocorrelation coefficient would have to exceed $|0.18|$ to be statistically significant at traditional levels of significance. Of the 108 autocorrelations of $s_{t+1} - s_t$, twelve lags for each of nine exchange rates reported in Fama's [59] Table 1, only two exceed 0.18. In contrast, the forward premium, $g_t - s_t$, is highly positively serially correlated for all currencies with first order autocorrelation coefficients ranging between 0.65 and 0.87 and with some remaining as large as 0.2 at lag 12.

Clearly, application of the standard test for serial correlation to the raw series, $s_{t+1} - s_t$, leads one to conclude that these variables are serially uncorrelated, yet the serial correlation of $g_t - s_t$ and the significance of β_2 in estimation of (4.5) imply that there is a statistically significant serially correlated expected rate of change of exchange rates. Presumably, the variance of the forecast error v_{t+1} in (4.13) is much larger than the variance of the expected rate of change of the exchange rate. This complicates inference because the signal to noise ratio is so high. Since any remaining residual serial correlation in the error term in (4.5) will also not be detectable with the standard test for autocorrelation, Fama's [59] assumption that this error is not serially correlated is probably quite reasonable.[25]

---

[25] Fama's [59] maintained assumption of conditional homoscedasticity is much less tenable. In light of Hsieh's [111] findings with weekly data discussed in Section 3, it may be that allowing for conditional homoscedasticity would actually strengthen Fama's findings by reducing his standard errors. This could change the interpretation of his results.

TABLE VI
SUR regressions from Fama [59]

$$g_t - s_{t+1} = \hat{\alpha}_1 + \hat{\beta}_1(g_t - s_t) + \hat{\varepsilon}^1_{t+1}, \quad s_{t+1} - s_t = \hat{\alpha}_2 + \beta_2(g_t - s_t) + \hat{\varepsilon}^2_{t+1}.$$

Part A: Unconstrained

Currency	$\hat{\alpha}_2(=-\hat{\alpha}_1)$	$\hat{\beta}_2(=1-\hat{\beta}_1)$	$\sigma(\hat{\alpha})$	$\sigma(\hat{\beta})$
Belgian franc	−0.36	−0.72	0.28	0.24
Canadian dollar	−0.26	−1.04	0.11	0.59
French franc	−0.48	−0.21	0.28	0.30
Italian lira	−1.08	−0.44	0.32	0.24
Japanese yen	0.12	−0.28	0.28	0.35
Dutch guilder	0.10	−0.78	0.27	0.25
Swiss franc	0.81	−1.15	0.42	0.50
UK pound	0.52	−0.69	0.26	0.51
Deutsche mark	0.23	−0.89	0.29	0.32

F tests 1. All β_2(or β_1) equal $F = 0.73$ P level $= 0.66$

         2. All α_2(or α_1) equal $F = 5.14$ P level $= 0.0001$

         3. All $\beta_2 = 0.0$(or $\beta_1 = 1.0$) $F = 2.81$ P level $= 0.003$

Part B: Constrained

$s_{t+1} - s_t =$

$\hat{\alpha}_B$	$+$	$\hat{\alpha}_C$	$+$	$\hat{\alpha}_F$	$+$	$\hat{\alpha}_I$	$+$	$\hat{\alpha}_J$	$+$	$\hat{\alpha}_N$	$+$	$\hat{\alpha}_S$	$+$	$\hat{\alpha}_{UK}$	$+$	$\hat{\alpha}_{WG}$	$+$	$\hat{\beta}_2(g_t - s_t)$
−0.34		−0.22		−0.57		−1.20		0.17		0.07		0.54		−0.49		0.14		−0.58
(0.28)		(0.10)		(0.27)		(0.27)		(0.28)		(0.27)		(0.35)		(0.23)		(0.28)		(0.13)

F test All α equal $F = 5.68$ P level $= 0.0001$

Note: The SUR regressions are completely complementary; that is, the intercepts in the $g_t - s_{t+1}$ and $s_{t+1} - s_t$ regressions sum to 0.0, the slopes sum to 1.0, and the residuals sum to 0.0 period-by-period. The subscripts on the $\hat{\alpha}$ coefficients in the constrained $s_{t+1} - s_t$ regressions are associated with dummy variables for the respective currencies.

Hodrick and Srivastava [108] implement a GMM estimator that does not impose the assumption of no serial correlation and basically confirm Fama's [59] statistical findings. It is to these findings that the discussion now turns.

Fama [59] performs seemingly unrelated regression (SUR) on the system of nine equations given by (4.5) for the nine countries in his study. His results are presented in Table VI which reproduces Fama's [59] Table 4. Notice that in each case the estimated $\hat{\beta}_2$ is negative, and it is statistically significantly negative, at least at the one percent marginal level of significance, except for the cases of

France, Italy, and Japan. The test that constrains all β_2 to be equal does not indicate any evidence against this hypothesis, while the hypothesis that all of the constants are equal is rejected at the 0.0001 marginal level of significance. Hence, Fama [59] constrains β_2 to be equal across countries, and his constrained estimate of β_2 is -0.58 with a standard error of 0.13.[26]

As noted above, such a finding carries with it the implication that the covariance between the risk premium and the expected rate of depreciation of the dollar relative to each of the nine currencies is negative and larger in absolute value than the variance of the expected relative rate of depreciation. Before turning to economic interpretations of these findings, consider Fama's [59] results for the two 61 month subperiods.

His SUR findings are presented in Table VII which reproduces Fama's [59] Table 7. Notice that each of the estimates of β_2 is negative in each subperiod except for Japan in the first subperiod and France in the second subperiod, but now only two countries have statistically significantly negative β_2's in the first subperiod (Belgium and Germany), and six countries have statistically significantly negative β_2's in the second subperiod (Canada, Italy, Japan, Netherlands, Switzerland and United Kingdom). Also, the joint test that all the slope coefficients are equal is rejected in the first subperiod at the 0.0095 marginal level of significance and at the 0.0016 level in the second subperiod.[27]

It seems quite surprising that in each subperiod there is evidence against the hypothesis that all slope coefficients are equal, yet there

---

[26] Longworth, Boothe, and Clinton [140] report results for an ordinary least squares specification like Fama's [59] of the rate of depreciation on a constant and the forward premium for the US dollar values of the Canadian dollar, for a sample period from July 1970 to December 1981, and for the US dollar values of the French franc, the Deutsche mark, the Japanese yen, and the UK pound, for a sample period from June 1973 to December 1981. They report results for one, three, six and twelve month maturities. For the Canadian one month rate the estimated beta was -0.389 with a standard error of (0.550). The estimated betas for other currencies were generally negative at all maturities although the beta for the Canadian dollar increased monotonically toward and past one as the forecast horizon increased.

[27] The F-statistics for these tests are not strictly appropriate since the right-hand-side variables are not strictly exogenous. Exact F-statistics would require normality of the error term and strict exogeneity of the right-hand-side variables. Nevertheless, Theil [208] argues that the F-statistic may actually be preferred since it is more conservative than the chi-square statistics discussed in Section 3.

TABLE VII
SUR regressions for 61-month subperiods from Fama [59]

$$g_t - s_{t+1} = \hat{\alpha}_1 + \hat{\beta}_1(g_t - s_t) + \hat{\varepsilon}^1_{t+1}, \quad s_{t+1} - s_t = \hat{\alpha}_2 + \hat{\beta}_2(g_t - s_t) + \hat{\varepsilon}^2_{t+1}.$$

Currency	Part A: Unconstrained			
	$\hat{\alpha}_2(= -\hat{\alpha}_1)$	$\hat{\beta}_2(= 1 - \hat{\beta}_1)$	$\sigma(\hat{\alpha})$	$\alpha(\hat{\beta})$
First subperiod: 8/31/73–4/7/78				
Belgian franc	0.00	−0.72	0.33	0.22
Canadian dollar	−0.22	−0.01	0.15	0.71
French franc	0.37	−0.45	0.41	0.56
Italian lira	−1.04	−0.39	0.52	0.37
Japanese yen	0.35	0.24	0.29	0.33
Dutch guilder	0.31	−0.53	0.34	0.31
Swiss franc	0.78	−0.41	0.40	0.73
UK pound	−0.54	−0.16	0.45	0.79
Deutsche mark	0.57	−2.23	0.35	0.59

F tests 1. All β_2(or β_1) equal $F = 2.56$ P level = 0.0095

        2. All $\beta_2 = 0$(or $\beta_1 = 1$) $F = 3.22$ P level = 0.0009

        3. All α_2(or α_1) equal $F = 3.92$ P level = 0.0002

Second subperiod: 5/5/78–12/10/82				
Belgian franc	−0.71	−0.41	0.45	0.41
Canadian dollar	−0.24	−1.78	0.16	0.82
French franc	−0.68	0.24	0.45	0.32
Italian lira	−1.11	−0.52	0.41	0.22
Japanese yen	1.08	−2.32	0.78	1.15
Dutch guilder	−0.03	−1.03	0.44	0.37
Swiss franc	1.87	−2.71	1.09	1.23
UK pound	−0.37	−3.06	0.35	0.78
Deutsche mark	−0.11	−0.46	0.48	0.40

F tests 1. All β_2(or β_1) equal $F = 3.17$ P level = 0.0016

        2. All $\beta_2 = 0$(or $\beta_1 = 1$) $F = 4.20$ P level = 0.0001

        3. All α_2(or α_1) equal $F = 3.92$ P level = 0.0002

Note: See Table VI.

is very little evidence against this hypothesis when parameters are estimated from the full sample. Statistical intuition suggests that standard errors will be larger in smaller samples, and consequently it should be more difficult to reject hypotheses in shorter data sets. Just the opposite is true here, which may be an indication that extreme points in the data are exerting more influence in the shorter samples than in a longer one. Perhaps each sample has an extreme observation, but they are opposite in sign. In any case, as Fama [59] notes, the general impression of negative covariation between the risk premium and the expected relative rate of depreciation is generally preserved in the subsamples.

Fama [59] finds the empirical results of his study to be somewhat troublesome. He states (p. 327), 'A good story for negative covariation between p_t and $E_t(s_{t+1} - s_t)$ is difficult to tell'. He suggests that a plausible story for negative covariation might be constructed from the Lucas [145] model discussed in Section 2, although he does not pursue the argument formally.

4.1a. Consistency of negative covariation and theory

Hodrick and Srivastava [108] do investigate whether negative covariation is a plausible outcome of the Lucas [145] model.[28] They also argue that, at a purely intuitive level, negative covariation between p_t and $E_t(s_{t+1} - s_t)$ is what might be expected. Their reasoning is that p_t is the expected return to selling foreign currency forward while $(-p_t)$ is the expected return from buying foreign currency forward and reselling the foreign currency for dollars in the spot market. Hence, $(-p_t)$ is a dollar denominated return and ought to increase with expected inflation in the US. Consequently, when something like expected inflation drives up the expected rate of depreciation of the dollar relative to foreign currencies, the expected dollar profit in the forward market $(-p_t)$ must also increase. This creates negative covariation between p_t and $E_t(s_{t+1} - s_t)$.

This intuitive reasoning is also supported by an examination of the determination of the risk premium and the expected rate of depreciation in the Lucas [145] model. From (2.5) and the definitions of the intertemporal marginal rates of substitution notice that

$$E_t[(S_{t+1} - S_t)/S_t] = E_t(Q^n_{t+1,1}/Q^m_{t+1,1}) - 1 \qquad (4.14)$$

where $Q^m_{t+1,1}$ is defined in (2.7) and $Q^n_{t+1,1}$ is defined in (2.10). A

---

[28] Adams and Boyer [1] also find negative covariation quite plausible in a monetary model of exchange rate determination with exogenous risk premiums.

similar deriation of the risk premium gives

$$E_t[(G_{t,1} - S_{t+1})/S_t] = E_t(Q^n_{t+1,1})/E_t(Q^m_{t+1,1}) - E_t(Q^n_{t+1,1}/Q^m_{t+1,1}).$$
(4.15)

Consequently, the covariance between the risk premium and the expected relative rate of depreciation using (4.14) and (4.15) is

$$C\left[\frac{G_{t,1} - E_t(S_{t+1})}{S_t} ; \frac{E_t(S_{t+1}) - S_t}{S_t}\right]$$

$$= C\left[\frac{E_t(Q^n_{t+1,1})}{E_t(Q^m_{t+1,1})} ; E_t\left(\frac{Q^n_{t+1,1}}{Q^m_{t+1,1}}\right)\right] - V\left[E_t\left(\frac{Q^n_{t+1,1}}{Q^m_{t+1,1}}\right)\right].$$
(4.16)

Since it is not possible to sign the covariance on the right-hand side of (4.16), Hodrick and Srivastava [108] examine a Taylor's series approximation of (4.16) and an example economy. Since the Taylor's series approximation also does not provide a completely satisfactory experiment, consider the example economy.

Let the period t utility function in (2.1) be given by the Cobb–Douglas specification, $U(X_t, Y_t) = AX^\alpha_t Y^{(1-\alpha)}_t$, which is evaluated at the equilibrium consumption levels. Then, the marginal utilities with respect to X and Y at time t are $U^x_t = \alpha AX^{(\alpha-1)}_t Y^{(1-\alpha)}_t$ and $U^y_t = (1 - \alpha)AX^\alpha_t Y^{(-\alpha)}_t$. Hodrick and Srivastava [108] follow Domowitz and Hakkio [42] in assuming that X_{t+1}, Y_{t+1}, M_{t+1}, and N_{t+1} are conditionally log normally distributed and that the variables are not correlated contemporaneously. Letting lower case letters indicate natural logarithms of their upper case counterparts, the distributional assumption is

$$x_t \sim N(E_{t-1}(x_t), \sigma^2_{xt}),$$ (4.17a)

$$y_t \sim N(E_{t-1}(y_t), \sigma^2_{yt}),$$ (4.17b)

$$m_t \sim N(E_{t-1}(m_t), \sigma^2_{mt}),$$ (4.17c)

and

$$n_t \sim N(E_{t-1}(n_t), \sigma^2_{nt}).$$ (4.17d)

Using these distributional assumptions and the assumed utility function, the expected relative rate of depreciation in (4.14) is

$$E_t(Q^n_{t+1,1}/Q^m_{t+1,1}) - 1 = \exp\{E_t m_{t+1} - E_t n_{t+1}$$
$$+ n_t - m_t + (1/2)(\sigma^2_{mt+1} + \sigma^2_{nt+1})\} - 1$$ (4.18)

and the risk premium from (4.15) is

$$- \exp\{E_t m_{t+1} - E_t n_{t+1} + n_t - m_t$$
$$+ (1/2)(\sigma^2_{mt+1} + \sigma^2_{nt+1})\}[1 - \exp(-\sigma^2_{mt+1})]. \quad (4.19)$$

Since (4.18) and (4.19) are determined by the same six variables, $\{n_t, m_t, E_t n_{t+1}, E_t m_{t+1}, \sigma^2_{nt+1}, \sigma^2_{mt+1}\}$, and because the partial effect of any of these variables is opposite in sign for the expression in (4.18) versus the expression in (4.19), the covariance between the risk premium and the expected relative rate of depreciation must be negative. Thus, negative covariation, per se, does not appear to be a puzzle.

The apparent puzzle is why the negative covariation is so large. If we assume that the statistical time series properties of the data satisfy the assumptions of stationarity and ergodicity, and we assume that a sample of ten years of monthly data is large, the statistical analysis tells us that the variability of the risk premium is sufficient to make the forward premium predict the wrong direction for the expected rate of change of the exchange rate. This finding is considered to be quite counterintuitive by many economists.[29]

Fama [59] notes that if market inefficiency due to inexperience with flexible exchange rates is viewed as the cause of this apparently perverse prediction, then the subsample results indicate that agents did not learn to correct their predictions in the second five years of the sample after having experienced negative prediction errors in the first five years.

---

[29] Gregory and McCurdy [89] argue that the single equation (4.5) is an inappropriate vehicle to investigate the unbiasedness hypothesis because the alternative hypothesis of a time varying risk premium is not fully specified. Their argument is easily extended to criticize Fama's [59] approach since there is no strong reason to assume existence of the unconditional variance and covariance that are presumed to be measured by the estimation. It may be that only conditional moments exist and that these moments are functions of various government policies. Then, (4.5) is subject to a type of Lucas [142] critique since $\hat{\beta}$ depends on the sample and is not constant from one time period to another if government policies change. An example of this might be the general weakness of the dollar during the Carter administration which was one force that prompted a change in operating procedures at the Federal Reserve in 1979. Such events may be evidence of stochastic process switching; see Flood and Garber [64]. Frankel [68] also argues that risk aversion cannot explain the empirical magnitudes described in this section. His argument is based on conditional homoscedasticity and is addressed by Giovannini and Jorion [86] who find no inconsistency.

4.2. Analysis of unbiasedness in real terms

In the derivation of the risk premium and the discussion of unbiasedness in Section 3, it is noted that risk neutrality, *per se*, implies (3.31). Engel (1984) defines the real profit on a sale of foreign currency as

$$e_{t+1} = [(G_t - S_{t+1})\pi_{t+1}^m]$$ (4.20)

in which case (3.1) implies

$$E_t(e_{t+1}) = 0.$$ (4.21)

By taking a stand on how to measure the purchasing power of the dollar, π_{t+1}^m, Engel [54] obtains an estimable specification of the risk neutrality hypothesis, or the absence of expected real profits. The proposition (4.21) implies that e_{t+1} is uncorrelated with all information in the time t information set. Engel [54] recognizes that (4.21) can be true, yet tests of unbiasedness such as those discussed in Section 3 can reject that hypothesis because of time variation in the conditional covariance in (3.32). Therefore, Engel [54] specifies two sets of tests of the proposition (4.21). One set of tests is a weak-form test in which e_{t+1} is regressed on four lagged values of itself. The other set of tests regresses e_{t+1}^j for a particular currency on e_t^i for four other currencies as well as one own lag, e_t^j.

Engel [54] also defines the variable $u_{t+1} = g_t - s_{t+1}$ and performs the same two sets of tests described above except with u_{t+1} everywhere replacing e_{t+1}. These tests amount to examination of the unbiasedness hypothesis in its logarithmic representation.

The data for the study are US dollar values of the Canadian dollar, the French franc, the Deutsche mark, the Japanese yen, and the UK pound. The sample period is monthly observations from October 1973 to April 1982. Two sets of price indexes are employed to measure the purchasing power of the dollar. The first is the US consumer price index. The second set consists of a constructed price index appropriate for an 'average resident' in the world. It is a weighted geometric average of consumer prices for the six countries (Canada, France, Germany, Japan, the United Kingdom, and the United States) with weights equal to the countries' relative shares in their composite GNP. The five non-dollar price indexes are first converted into dollar terms by multiplying them by the dollar exchange rate prior to their inclusion in the index.

The estimation technique is ordinary least squares with the

heteroscedastic consistent covariance matrix given in (3.43) utilized when appropriate. In contrast to Hsieh [111], Engel [54] only employs the heteroscedastic correction in cases in which the residuals from the OLS regression fail the test for heteroscedasticity proposed by White and described above.

Surprisingly, given the previous results, for the first set of tests, only the Japanese yen residuals fail the heteroscedasticity test. This is true of both specifications of e_{t+1} and for the equations with u_{t+1}. Also somewhat surprising is Engel's [54] finding that the null hypothesis of zero coefficients cannot be rejected in any of the three specifications. The chi-square statistics with five degrees of freedom are never larger than 7.1, and the critical level corresponding to the 0.05 marginal level of significance is 11.1.

For the second set of tests, the OLS residuals for both specifications of e_{t+1} fail the heteroscedasticity test, except for the Canadian dollar. None of the u_{t+1} regression residuals fail the White test. Also, there again is no evidence that the coefficients in any of the equations are nonzero. The largest chi-square statistic with six degrees of freedom is 8.81, and the 0.05 critical level is 12.6.

The results from Engel's [54] study are seemingly at odds with Fama's [59] analysis. Although the data are not from the same source, the sample period is almost identical. Both studies use monthly data from 1973 to 1982. Fama's sample contains nine additional observations which is less than ten percent additional data.

Two factors appear to be driving the differences in statistical inference. First, Fama utilizes a specification that employs the forward premium as the only stochastic regressor, and it has a much smaller variance than the lagged dependent variable, u_t. Consequently, the a priori likelihood of being able to explain much of the variability in u_{t+1} is limited in Fama's [59] study, but if market efficiency in the forward foreign exchange markets is characterized by a time varying risk premium, the part of u_{t+1} that can be explained with time t information ought to be relatively small if movements in risk premiums are small. If there is serial correlation in the u_t series, it is likely to be weak because of the large unanticipated change in exchange rates. Since the forward premium is explicitly forward looking in the sense of the decomposition in (4.4), its variance is not overlaid with extraneous noise, and the true signal is more likely to appear. This may be particularly important in environments without constant serial correlation coefficients.[30]

The second factor that is different between the two studies is the estimation technique. Fama [59] uses traditional OLS and SUR, both of which provide strong rejections of unbiasedness conditional on the appropriateness of their standard errors, and Engel's heteroscedasticity tests actually support Fama's statistics. Engel [54] uses OLS with the modified standard errors, where appropriate, but he employs no system techniques which would allow tests of joint hypotheses.

It is my conjecture that the differences between the two studies arise more from the use of different regressors than from the alternative statistical techniques. Nevertheless, Engel's [54] emphasis on the necessity of conducting the tests in an appropriate specification is a point worth reemphasizing. The only potential problem with the analysis is that the quality of the data on price indexes may not be particularly good compared to the point in time asset price data. These issues arise again in the discussion of tests of formal models of risk premiums which are the topics of Section 5.

4.3. Contentions of market inefficiency

Bilson [14] investigates what he refers to as the speculative efficiency hypothesis. His econometric analysis is essentially an investigation of the unbiasedness hypothesis which is a precursor of

---

[30] The results in Rose and Selody [184] might also be explained by their strict use of forecast errors. They use 2516 daily observations on US dollar-Canadian dollar exchange rates from 1971 to 1980 to investigate the logarithmic specification of the unbiasedness hypothesis. Logarithmic forecast errors or ex post logarithmic returns for ten different horizons are regressed on one lag of each of the ten returns and five lags of the return under consideration. The ten returns with marginal levels of significance in parenthesis are the one (0.04), two (0.17), three (0.22), six (0.42), and twelve month (0.44) pure forward contracts and the rollover strategies of a two-month contract covered in one month (0.03), the three-month contract covered in one month with a two-month contract (0.03), the three-month contract covered in two months with a one-month contract (0.13), the six-month contract covered in three months with a three-month contract (0.29) and the twelve-month contract covered in six months with a six-month contract (0.27). The marginal levels of significance correspond to the value of the chi-square statistic with sixteen degrees of freedom that tests the hypothesis that all coefficients in a regression are zero. Rose and Selody [184] argue that the evidence of the tests (p. 672) "has not indicated the presence of an economically interesting inefficiency."

Fama's [59] specification discussed above. The novel aspect of Bilson's analysis is that he goes a step further than has typically been the case in studies of unbiasedness to ask whether the trade-off between risk and return on a trading strategy implied by the parameter estimates and the rejection of unbiasedness is consistent with the types of trade-offs found in other asset markets.

Based on an out-of-sample investigation of the relationship between the expected return on a portfolio of positions in the forward market and the variability of the payoffs on these positions, Bilson [14, p. 449] concludes, "the profit/risk ratio appears to be too large to be accounted for in terms of risk aversion." To understand how he was led to this conclusion, consider Bilson's investigation of unbiasedness.

Bilson's [14] specification of the test of unbiasedness is

$$s_{t+1}^i - g_t^i = \delta_1(g_t^i - s_t^i)^S + \delta_2(g_t^i - s_t^i)^L + \varepsilon_{t+1}^i \qquad (4.20)$$

where the superscripts (S and L) indicate values of the forward premium expressed at an annual rate that are smaller or larger than 10% in absolute value, respectively. The data for his study consist of monthly observations (a 4 week interval) on the nine currencies available from the Harris Bank Data Base as described above in the discussion of Fama [59]. The initial sample period was July 1974 to January 1980. Bilson's estimates of δ_1 and δ_2 are constrained to be the same across currencies, and he also employs the SUR technique. His parameter estimates with their standard errors in parenthesis are $\hat{\delta}_1 = -0.741$ (0.15) and $\hat{\delta}_2 = -1.280$ (0.12). In terms of the parameter β_2 in (4.5), $\hat{\beta}_2 = 1 + \hat{\delta}_i$, $i = 1, 2$, and we find that the negative slope coefficient in Fama's [59] analysis appears to be driven by the observations associated with values of the forward premium that are large in absolute value. It is still possible to reject the hypothesis that $\beta_2 = 1$, but as Bilson notes, the large values of the forward premium appear to be the ones that are associated with predictions of relative depreciation that are the wrong direction. This suggests that small sample problems may plague these studies since one explanation of these findings involves stochastic process switching that is insufficiently anticipated or improperly represented during the sample compared to a priori possibilities. These issues are discussed in more detail below.

Now consider Bilson's [14] profitability argument. Based on

estimates of (4.20), Bilson forms a vector of out-of-sample expected profits, r_t, for the nine currencies with typical element $r_t^i = \delta_{1t}(g_t^i - s_t^i)^S + \delta_{2t}(g_t^i - s_t^i)^L$. The coefficients now have t subscripts to indicate that they are based on information up to that time. The estimated covariance matrix of the contemporaneous errors in (4.20), denoted Ω_t to indicate dependence on sample data up to time t, is then used in combination with r_t to form a portfolio with dollar value positions in the forward market for nine currencies denoted by q_t. The portfolio weights are chosen in the following way:

$$\min_{q_t} q_t'\Omega_t q_t \text{ subject to } q_t'r_t = \pi^*, \qquad (4.22)$$

where π^* is a desired target profit. The solution to (4.22) is

$$q_t^* = \Omega_t^{-1}r_t(r_t'\Omega_t^{-1}r_t)^{-1}\pi^*. \qquad (4.23)$$

Equation (4.23) demonstrates that the optimal positions are proportional to the desired profit with the factor of proportionality depending upon the covariance matrix of the profits and the vector of expected profits.

As Bilson notes, the 'efficient frontier,' defined as the locus of points of maximum expected profit for a given standard deviation of profit, is linear in this case,

$$\pi^* = k_t\sigma(\pi^*), \qquad (4.24)$$

where the factor of proportionality is $k_t = (r_t'\Omega_t r_t)$. The efficient frontier is a linear ray through the origin because the speculator can avoid both profit and risk by taking zero investment in all foreign currency forward markets and because the system (4.23) is linearly homogeneous in its nominal magnitudes.

Hodrick and Srivastava [106] note that the specification of the portfolio problem presumes that, in contrast to the rational investor of Section 2, the speculator in Bilson's framework cares only about the first two moments of the profit on his forward market portfolio and not about the covariation of the profit with the returns on other assets or with his consumption stream. In this sense it is necessary to find a trade-off between risk and expected return that is too good to be consistent with risk aversion in order to conclude that the forward market is inefficient. Merely demonstrating that some trade-off exists may simply be a reflection of the rejection of the unbiasedness hypothesis and is consistent with the existence of a risk premium.

The basis of Bilson's [14] contention that the market is inefficient is an examination of standardized expected profits (SRE), which are defined to be expected profits divided by the standard deviation of the portfolio, and standardized actual profits (SRA), which are analogously defined using actual profits.[31] Although the estimation is conducted with nonoverlapping data, the out-of-sample experiment is conducted with weekly data from February 1980 to January 1981. The SRE's range from 0.48 to 1.85 with an average of 0.929. The SRA's ranged from -2.36 to 3.29 with an average of 0.857. If profit on the portfolio is being drawn from a normal distribution, SRE provides the interpretation that profits are expected to be unity with a standard deviation of the inverse of SRE. Hence, with ninety-five percent confidence, we would expect profit to be drawn from unity plus or minus two times the inverse of SRE.

Bilson [14] notes that an average SRE of 0.929 implies that if expected profits are one, the two standard deviation band runs from approximately negative one to positive three. This is indeed a favorable trade-off and prima facie evidence against market efficiency. Bilson investigates whether the trade-off is stable over time. Since actual profits are not independent draws, due to the overlap of the data, the error structure between the difference of SRA and SRE forms a third order moving average process. Bilson allows for this process by iteratively estimating a regression of SRA–SRE on a constant and a time trend. The results indicate some relatively strong evidence against the hypothesis that SRE is an unbiased predictor of SRA since the constant term is 0.56 with a standard error of 0.30 while the coefficient of the time trend

---

[31] Geweke and Feige [82] examine a different type of risk-return tradeoff. They take risk to be measured by the unconditional standard deviation of the error term in a specification like (4.20). Return is assumed to be measured by the unconditional standard deviation of the explained part of the regression. Their specification also misses the point that variance of returns may be an inappropriate measure of the riskiness of an asset and may not be related to its expected returns. Salemi [187] uses an autoregression with time varying parameters to forecast the future spot rate. His measure of risk is the uncertainty in the forecast which is composed of three components, the true forecast error, the sampling error in the estimated coefficient and the parameter uncertainty of the changing coefficients. He finds a relation between the expected return on the forward contract and the measure of risk.

is -0.24 with a standard error of 0.09. Bilson concludes from the signs of the coefficients that significant predictable speculative profits had been available during his out-of-sample experiment, but that they may have been arbitraged away by the end of the time period. Nevertheless, he also concludes that the average SRE indicates that the market is inefficient.

Hodrick and Srivastava [106] perform two alternative experiments of trading strategies analogous to Bilson's [14] in order to address the issue of the size and volatility of trading profits. Their data are monthly nonoverlapping observations on spot exchange rates and one month forward rates from July 1973 to September 1982. The five exchange rates in their study are the US dollar values of the French franc, the Japanese yen, the Swiss franc, the UK pound, and the Deutsche mark. One goal of the analysis in Hodrick and Srivastava [106] is to investigate what the performance of Bilson's strategy would have been over several years if someone had begun using it in the early part of the sample.

Their first experiment consists of sequentially estimating and simulating the trading strategy using the following model to generate expected profits:

$$(S_{t+1}^i - G_t^i)/S_t^i = \beta_{i0} + \sum_{j=1}^{5} \beta_{ij}(G_t^j - S_t^j)/S_t^j + u_{t+1}^i, \qquad i = 1, \ldots, 5,$$

(4.25)

where superscripts index the five currencies. They use the first twenty-five observations to compute the coefficient vector, β_t, and the covariance matrix, Ω_t, which is estimated with maximum likelihood. Combining β_t with the next set of forward premiums provides an expected profit vector which is combined with Ω_t to form an optimal portfolio as the solution to (4.22). The procedure of OLS estimation and formation of portfolios is then repeated by adding an observation at each date until the end of the sample period.

The second experiment allows for stochastic parameter variation through time. It assumes that the coefficient vector follows a first-order autoregressive process,

$$\beta_t = A\beta_{t-1} + \varepsilon_t,$$

(4.26)

where the updated coefficients are given by the Kalman filtering formula:

$$\beta_t = A\beta_{t-1} + (AP_t x_t')(x_t P_t x_t' + \Omega_t)^{-1}(y_t - x_t A\beta_{t-1}). \quad (4.27)$$

Here x_t is the vector of right-hand-side variables in (4.25), and y_t is the vector of dependent variables in (4.25). The covariance matrix of β_t is

$$P_t = AP_{t-1}A' + Q_t$$

where Q_t is the covariance matrix of ε_t. In order to run the experiments, values for A, β_0, P_0, Q_t and Ω_t have to be specified a priori. The prior on the coefficients, β_0, is specified to be the OLS estimate of β based on the first 24 periods. The prior covariance matrix, P_0, is specified the same way. Since they did not have a well defined prior on Q_t, it is specified to produce P_t equal to P_0 for all time by setting $Q_t = P_{t-1} - AP_{t-1}A'$. The covariance matrix Ω_t is again specified by maximum likelihood estimation as in the first experiment, and the matrix A is specified to be 0.75 times the identity matrix.

The two experiments are run over 83 nonoverlapping monthly observations. Hodrick and Srivastava [106] propose a test of whether profits at time t are drawn from a normal distribution with mean π^* and variance σ_t^2, where σ_t^2 is the estimated portfolio variance at t. They argue that since σ_t^2 is estimated, $(\pi_t - \pi^*)/\sigma_t$ has a t-distribution. This is correct, but the degrees of freedom are not $(t - 6)$ as is argued. This would be correct if only one equation were estimated with t observations and 6 parameters, but with five equations and thirty free parameters, a more appropriate measure would be $(5t - 30)$. Since the t-distribution approaches the normal as the degrees of freedom become large, an alternative way of discussing the adequacy of the results considers actual profit minus expected profit divided by the standard deviation of the portfolio to be a standard normal random variable. In this case the average SRA minus the average SRE should be distributed as a normal random variable with mean zero and variance equal to one divided by the number of out-of-sample experiments.

Hodrick and Srivastava [106] report for the first experiment that the average SRA is 0.871 while the average SRE is 0.211. With 83 out-of-sample nonoverlapping observations, the difference between the two averages of 0.660 is 6.01 standard deviations from zero. Consequently, the first experiment does not produce actual profits and variances of profits from the distribution that is anticipated. In contrast, the second experiment produces an average SRA of 0.660 and an average SRE of 0.620. With 83 observations the difference of the averages of 0.04 is 0.36 standard deviations from its expected value. Even though use of the standard normal distribution is only justified in large samples, one can argue that the deviation in the second experiment is not very economically significant.

Hence, since the hypothesis that $\pi^* = 1$ in the second experiment is not rejected, it makes sense to examine the implied risk-return trade-off as measured by the mean and standard deviation of profits. Applying the two standard deviation band to the average SRE implies that if expected profits are one, the ninety-five percent confidence interval is -2.03 to 4.03. This represents a less favorable trade-off than Bilson [14] reports. Nevertheless, it is reasonably favorable.

One standard of comparison is the SRA generated from the Standard and Poor's 500. Ibbotson and Sinquefield [115] report an average rate of return on this portfolio from 1926 to 1981 of 0.091 and a standard deviation of return of 0.212. These data produce an SRA for the portfolio of 0.429. The average SRA for the foreign currency portfolio is not terribly larger than this. Hodrick and Srivastava [106] also report the range of SRE. When SRE is at its minimum of 0.061, the implied ninety-five percent confidence interval is -31.787 to 33.787 which is extremely unfavorable. When the SRE is at its maximum of 4.573, the two standard deviation band is from 0.563 to 1.437, which implies an almost sure profit.

It is probably unwise to place much emphasis on the extreme values given that they represent values from estimated parameters. Hence, Hodrick and Srivastava [106] conclude that the risk-return trade-off appears to be quite volatile and highly sensitive to the model used to form expected profits and variances of profits.

Bilson and Hsieh [15] note that neither of these early studies discussed above takes account of transactions costs which will lower perceived profitability, nor do they allow the speculator to increase his stake in the market when speculation appears favorable while reducing the stake when it is unfavorable. They also argue that the covariance matrix of the profits should be allowed to evolve through time and be more dependent on the present than is allowed by the maximum likelihood estimator. Unfortunately, in my view, they compute the covariance matrix of forecast errors only from the past 16 forecast errors. This places great importance on the previous year of data at the total exclusion of observations in the more distant past. During relatively tranquil years, this will probably understate the true conditional variance which distorts the value of the experiments.

Longworth, Boothe and Clinton [140] explore the out-of-sample profitability of several forecasting equations. Their strategy is similar to that described above. One of their specifications uses the rate of depreciation as the endogenous variable with a constant and the forward premium as the regressors. They estimate the equation over an initial sample of approximately three years of monthly data, form a forecast, and sequentially reestimate and form forecasts. When the specification predicts that the expected spot rate is greater than the forward rate, they take a long position of one unit of foreign exchange. When the reverse is true, they sell forward one unit of foreign exchange.

The mean profit in the case of Canadian dollar–US dollar exchange rate is 0.00209 with a standard error of 0.01379. Longworth, Boothe and Clinton [140] assume that the profits are drawn from a normal distribution with constant parameters.

Under such an assumption and with 95 observations, the probability that the mean profit is greater than zero can be calculated from the standard t-statistic. In this case the marginal level of significance is 0.069. Other forecasting rules produce comparable or lower values of the level of significance. Of course, if the distribution is not constant as was argued above, these levels of significance are probably not accurate reflections of the true statistics. The conclusion of Longworth, Boothe and Clinton [140], that almost certain evidence of unexploited speculative profit has been found, must be tempered. They also note that Boothe [18] demonstrates that taking account of transactions costs will also reduce the estimated profitability of the speculative profits although they argue that actual costs are so small as not to affect the profitability calculations to a significant extent.

4.4. Filter rule profitability

The previous section examined the profitability of a trading strategy based on forecasts of future spot exchange rates generated from regression models. This section considers simpler trading strategies known as filter rules that are based strictly on the past history of changes in exchange rates.

Alexander [6] pioneered the use of filter rules in analysis of stock price data. If the time series of equilibrium stock prices are well approximated by a model with a positive conditional expected return, a trading strategy based strictly on the past history of stock prices that suggests periods in which the stock should be sold short will not provide superior profit to a buy and hold strategy. Filter rules can also be thought of as attempts to test the profitability of trading strategies proposed by chartists. Chartists and proponents of inefficient markets often argue that prices are subject to dynamics induced by trading. One variant of the price dynamics viewpoint is the "bandwagon" hypothesis. Dooley and Shafer [45, p. 47] state,

According to this hypothesis a small set of market leaders are known, or thought, to have more accurate information concerning the factors that will affect future prices. When this set of market participants buys or sells, generating a price change, a signal is provided to other market participants to jump on the bandwagon. The followers are thought generally to overshoot the new equilibrium price.

The filter rule methodology is designed precisely to look for this overshooting which is characteristic of an inefficient market. In the context of foreign exchange markets, Dooley and Shafer's [45] version of an X percent filter rule works like this. An investor begins with no position in either currency. After an X percent depreciation of say the dollar relative to the Deutsche mark from some previous trough, the investor borrows dollars and invests in Deutsche marks until the dollar appreciates relative to the Deutsche mark by X percent. At this time the investor reverses his positions by borrowing Deutsche marks and lending dollars. At the end of the period of analysis, all loans are repaid and profits or losses are evaluated.

Dooley and Shafer [45] use overnight Eurocurrency interest rates as their investment and loan interest rates. They express profits and losses as annual rates of return on the size of the position. For example, if the speculator maintains a $1 short or long position

TABLE VIII
Filter rule profitability from Dooley and Shafer [45]

Currency	Sample period	Annual percentage profit of 1% filter	Var. $\times 10^{-5}$	Std. dev. profit
Belgian franc	1	10.17	3.88	8.86
	2	4.11	1.74	5.94
	3	8.78	5.19	10.25
Canadian dollar	1	−0.72	0.18	1.91
	2	2.93	0.60	3.49
	3	8.02	0.54	3.31
French franc	1	17.31	4.20	9.22
	2	−0.64	1.83	6.09
	3	12.13	4.52	9.57
Deutsche mark	1	5.71	4.85	9.91
	2	4.98	1.85	6.12
	3	10.74	5.00	10.06
Italian lira	1	6.51	N.A.	N.A.
	2	7.14	2.54	7.17
	3	4.81	3.60	8.54
Japanese yen	1	3.89	1.96	6.30
	2	15.45	2.04	6.43
	3	17.28	5.11	10.17
Dutch guilder	1	17.06	3.87	8.85
	2	4.24	1.68	5.83
	3	16.05	5.20	10.26
Swiss franc	1	10.44	6.24	11.24
	2	11.09	4.79	9.85
	3	8.19	6.45	11.43
UK pound	1	8.88	1.48	5.47
	2	11.12	2.28	6.79
	3	12.58	4.35	9.39

Notes: Sample period 1 is from 13 March 1973 to 5 September 1975. Sample periods 2 and 3 correspond to the first and second halves of the period from 8 September 1975 to 6 November 1981. Not available in the original is denoted N.A. Standard deviation of profit is calculated as $0.9(\text{Var})^{1/2}(250)^{1/2}(100)$.

throughout a year and earns a profit of \$0.10, he is said to have earned 10 percent per annum by following the trading rule. Transaction costs are captured by selling or investing at the bid price and by buying or borrowing at the ask price in each market.

The results of a one percent filter rule for three different sample periods are reported in Table VIII which is taken from Dooley and

Shafer's [45] Table 3-5.[32] The most striking feature of Table VIII is the consistent profitability of the filter across currencies. While this is not to be unexpected since the exchange rates are all dollar denominated, the correlation across currencies is far from one. Losses occur in only two subperiods, sample one for the Canadian dollar and sample two for the French franc.

One argument against the analysis of data with filter rules is always that sufficient search across alternatives will produce a profitable filter. Dooley and Shafer [44] first published the results of their study of sample one. Samples two and three do not demonstrate a noticeable decrease in profitability. In contrast, the profitability for four of the nine currencies is the largest in the third sample.

An interesting feature of Dooley and Shafer's [45] analysis is that they include three artificially constructed random walks in their analysis. This serves as a reminder that measured positive profitability could occur by chance (notice R.W. #3, sample 3), but the preponderance of positive profitability suggests that chance is an unlikely explanation of the results.

Sweeney [207] argues correctly that the absence of statistical tests of the significance of the profits from filter rule analyses and the lack of adjustment for an appropriate model of risk and return make interpretation of the results of filter rules difficult. Dooley and Shafer [45, p. 65] agree with the latter sentiment because they state

the profitability of the filter rules suggests that deviations from a martingale are important and that exchange markets are for many currencies either not efficient in use of price information or differences in rates of return were large and variable during the sample period.

Since Dooley and Shafer's [45] model of risk and return was the unbiasedness hypothesis that predicts a white noise process for the profits from borrowing dollars and investing in a foreign currency, it is possible to construct a crude test of the statistical significance of the filter rule profit.[33]

---

[32] Dooley and Shafer [45] also examine the profitability of three, five, ten, fifteen, twenty and twenty-five percent filter rules. The results for the three and five percent rules are roughly similar to those of the one percent rule. The larger rules produced several large losses.

[33] This is complicated because Dooley and Shafer [45] are unclear about what is done with interim profit. Since they are borrowing and investing in the overnight money markets, the daily profits or losses ought to be reinvested in calculating total profit. This complication is ignored here.

Dooley and Shafer [45] report the variance of daily changes in the natural logarithms of exchange rates in Table 3-5 which is reproduced in Table VIII. This overstates the variance of daily profit, expressed as a rate of return to borrowing a dollar, to the extent that it leaves out the expected change in exchange rates due to the interest differential. A rough adjustment can be made following the idea of Mussa [165] who argues that over 90 percent of changes in exchange rates are unanticipated, perhaps even a larger percentage with daily data. An adjustment is to use 0.9 times the standard deviation of daily changes in exchange rates as the daily standard deviation of profits. The daily standard deviation is also multiplied by the square root of 250 and by 100 to express it as an annual percentage profit.

Notice that in only three of the 27 separate cases is the annual percentage profit greater than two standard deviations from zero. Also, thirteen of the observations are within one standard deviation of zero.

Hence, even by the criterion of providing profit compared to the naive risk adjustment implied by the unbiasedness hypothesis, the filter rule profits of Dooley and Shafer [45] do not appear to be particularly significantly different from zero by this standard. The fact that almost all of the observations are positive, though, suggests that this approach may overstate the lack of statistical significance of the filter rule profits.

Sweeney [207] compares his filter rule profitability to a benchmark strategy of buying and holding the foreign currency investment and recognizes that buy and hold may require an expected return due to risk. He applies his analysis to the US dollar–Deutsche mark exchange market. Sweeney [207] interprets the filter rule discussed above in a slightly different way. After an appreciation of the Deutsche mark relative to the dollar of X percent, the US speculator invests a dollar in an overnight Deutsche mark denominated asset that pays the riskless Deutsche mark rate of return. The position is maintained until an X percent depreciation of the Deutsche mark relative to the dollar when the individual repatriates the funds and invests in the riskless dollar asset.

Sweeney [207] recognizes that the buy and hold strategy may be risky, and in such a situation, the unbiasedness hypothesis is an inappropriate characterization of the equilibrium risk-return tradeoff. He employs the static capital asset pricing model as his risk adjustment postulating as in (2.23) that the expected spot rate minus the forward rate divided by the current spot rate is equal to a beta times the excess expected return on the market portfolio over the risk-free return. He treats this required return as a constant denoted g. For a sample of N days, the average risk adjusted profit on buy and hold would provide an estimate of g. Let the sample average buy and hold return be denoted $\bar{R}_{BH}$.

If a filter rule indicates uncovered investment in the foreign currency asset for $(1 - f)$ percent of a sample, then the sample average excess return to the filter, $\bar{R}_F$, is the sum of excess profit on the days in the foreign currency divided by the total number of days because on days when the speculator is out of foreign currency he bears no risk. It has expected value $(1 - f)g$.

Sweeney [207] examines the statistic

$$X = \bar{R}_F - (1 - f)\bar{R}_{BH} \tag{4.28}$$

in order to determine whether filter rules beat buy and hold. Notice that X can be positive even if average filter rule returns are smaller than average buy and hold returns because the speculator bears no risk f percent of the time during the sample. The expected value of X is zero, and the variance of X, under the null hypothesis that daily excess returns are independently and identically distributed with variance σ_u^2, is

$$\sigma_x^2 = f(1 - f)\sigma_u^2/N. \tag{4.29}$$

Sweeney [207] examines the profitability of his filter rule for a sample of 1288 trading days between 1975 and 1980 for the US dollar–Deutsche mark investments. His Table 2 reports the X statistic in (4.28) and tests of its statistical significance using the standard deviation derived from (4.29) for various filters ranging from 0.5 percent to 10 percent.

The results for the one-percent filter indicate values after transactions costs that are statistically significantly different from zero at conventional levels. The value of X, net of transaction costs, is 3.88 expressed at an annual percentage rate. The standard deviation is 1.83.[34] All the other filters produce statistically insignificant profits after transaction costs.

---

[34] Since bid-ask data were unavailable, Sweeney [207] assumed transaction costs were one-eighth of one percent of asset value for each round trip transaction. There were 42 round trips in the 1289 trading days with the one percent filter.

Because of lack of data, Sweeney [207] also examines the profitability of the filter rules without adjusting for the interest differential for the Deutsche mark and a number of other currencies. While recognizing that this is theoretically unsound, he appeals to the findings in the case of the Deutsche mark that the test statistics with and without interest differentials are quite similar. Intuitively, this will be true if the average interest differential for days in the foreign currency is approximately equal to the average for the whole period.

Table IX reports the annual percentage excess returns from the one-percent filter rule for ten currencies and two sample periods. The results are taken from Sweeney's [207] Tables 3 and 4. Since

TABLE IX
Filter rule excess returns from Sweeney [207]

Currency	Sample period	Annual percentage excess return of 1% filter	t-statistic
Belgian franc	1	7.25	2.32
	2	4.25	2.43
Canadian dollar	1	1.25	1.67
	2	3.00	3.43
Deutsche mark	1	6.50	1.86
	2	3.50	2.00
French franc	1	8.50	2.62
	2	3.50	2.15
Italian lira	1	5.00	2.86
	2	3.75	2.14
Japanese yen	1	1.75	0.82
	2	6.75	3.18
Swiss franc	1	9.00	2.25
	2	5.75	2.42
Swedish krone	1	7.50	2.50
	2	3.75	2.31
Spanish peseta	1	-1.75	-1.27
	2	4.75	2.00
UK pound	1	3.75	1.89
	2	6.50	3.71

Notes: Sample 1 is the first 610 trading days after 1 April 1973. Sample 2 is the next 1220 trading days ending in 1980. Annual percentage profits were calculated by multiplying Sweeney's reported values by 250. The t-statistics are the ratio of X to its standard error.

almost all of the values are positive, and 15 of the 20 t-statistics are greater than two, it is likely that a joint test of the significance of the excess returns would have a very small marginal level of significance.

Sweeney [207] offers several potential explanations of the profitability of the filter rules. First, they can be interpreted as evidence against the static capital asset pricing model in which case they might be consistent with alternative explanations of risk and return. Second, they may be evidence of market inefficiency and insufficient speculative capital. Third, they may represent profits that are available to speculators because of central bank intervention which systematically loses money by leaning against the wind.

The next section explores evidence on alternative models of risk and return other than the unbiasedness hypothesis. An interesting challenge for these models is to see whether they explain the apparent profitability of the filter rules.

5. ECONOMETRIC MODELS OF RISK PREMIUMS

The previous section explored alternative interpretations of the rejection of the unbiasedness hypothesis. While market inefficiency is not a totally implausible hypothesis, a time-varying risk premium is perhaps a more likely explanation given the abundant evidence that average holding period returns on assets have varied greatly. Ibbotson and Sinquefield [114, 115] document the existence of large differences in the average holding period returns on a variety of assets. Most financial economists view these differences as a reflection of risk premiums that are potentially traceable to risk aversion of individual investors. If agents are risk averse, the asset market theory of exchange rate determination combined with the intertemporal asset pricing theory discussed in Section 2 suggests that risk premiums ought to characterize the forward market for foreign exchange.

In this section I examine alternative econometric models of the risk premium. First, models that are based strictly on the time series properties of spot and forward exchange rates and asset prices are considered. Then, models that employ other market fundamentals are examined.

5.1. Models with no market fundamentals

There is a long tradition in financial economics of developing tests of asset pricing models that utilize only the comparatively accurately measured returns on assets. Other data that are fundamental from an economic perspective are ignored. This section discusses several tests of asset pricing models that do not incorporate market fundamentals such as the outstanding stocks of assets. Models that incorporate market fundamentals are discussed in Section 5.2.

In Section 2 a conditional capital asset pricing model is derived in (2.23). This expression is repeated here for convenience as (5.1) since it forms the basis of the empirical models of this section. The representation of the expected normalized profit on a long position in the forward market is

$$E_t(S_{t+k} - G_{t,k})/S_t = \beta_t^G E_t(R^b_{t+k,k} - R_{t,k}) \tag{5.1}$$

where $\beta_t^G = C_t[(S_{t+1} - G_{t,k})/S_t; R^b_{t+k,k}]/V_t(R^b_{t+k,k})$ and $R^b_{t+k,k}$ is the k-period return on an appropriate benchmark portfolio. Expression (5.1) is a conditional capital asset pricing model, and there are various ways to give the model empirical content.

5.1a. An index model

The first empirical approach in this area is by Roll and Solnik [181] who build on the theoretical work of Solnik [198]. Solnik [198] develops one of the first theoretical models of international asset pricing. Since it is an early model, its assumptions are relatively strong, and they have come under criticism for their empirical validity. While Roll and Solnik [181] admit that the assumptions of the model lack realism, they consider the theoretical arguments that motivate the existence of risk premiums to be sufficiently compelling to engage in an empirical study derived from the theory.

In Solnik's [198] model, the expected 'extraordinary exchange return', which is the left-hand side of (5.1), for currency i in terms of the numeraire currency $N + 1$, can be written as

$$E_t(S^i_{t+k} - G^i_{t,k})/S^i_t = b_i \sum_{j=1}^{N} w_j E_t(S^j_{t+k} - G^j_{t,k})/S^j_t, \tag{5.2}$$

for $i = 1, \ldots, N$, where b_i depends on the covariance of the exchange rate change of currency i in terms of currency $N + 1$ with the index of changes for the other N exchange rates, and where the weights, the w_j's are the net values of capital of country j which is foreign owned as a proportion of the total value of world capital expressed in a common currency, (see Solnik [198, pp. 34–35]). Clearly, an analogous representation can also be derived from the general expression (5.1) by writing

$$E_t(R^b_{t+k,k} - R_{t,k}) = (\beta^{G,i}_t)^{-1} E_t(S^i_{t+k} - G^i_{t,k})/S^i_t, \qquad (5.3)$$

for $i = 1, \ldots, N$, then dividing each of the N equations in (5.3) by N, and finally summing the N equations. Performing this calculation gives

$$E_t(S^i_{t+k} - G^i_{t,k})/S^i_t = \beta^{G,i}_t \sum_{j=1}^{N} N^{-1}(\beta^{G,j}_t)^{-1} E_t(S^j_{t+k} - G^j_{t,k})/S^j_t. \qquad (5.4)$$

In order to implement their model empirically, Roll and Solnik [181] treat the w_j parameters in (5.2) as constants each equal to one seventh and use data from eight countries in which case $N = 7$. They also substitute realizations of the variables for their conditional expectations. While this produces an estimable equation, application of ordinary least squares is only correct if the unanticipated innovations are mutually orthogonal. When the innovations are correlated, as they almost surely are given that the exchange rates are measured in a common currency, OLS will produce biased estimates of $\beta^{G,i}_t$ even if the cross equation constraints on the observations in (5.4) are employed in an iterative technique. Since the estimation technique seems to tell us nothing more than the fact that correlation exists in unanticipated changes in exchange rates measured in a common currency, the results of Roll and Solnik [181] are not discussed further.

5.1b. The static CAPM

Another early empirical study of the risk premium in the foreign exchange market is that of Robichek and Eaker [176]. They employ a static capital asset pricing model to price foreign exchange risk. Their asset pricing argument is developed in the following way. The present value of a sure claim to one unit of foreign currency in one period is the forward exchange rate divided by one plus the sure rate of interest or G_t/R_t in the notation employed here. The title to one unit of foreign currency to be purchased in the future spot market is a risky asset. The capital asset pricing model requires that the expected value of a future cash flow be discounted by one plus the appropriate risk adjusted rate of return. Let this return be denoted $E_t(R_{t+1}^s)$. Since the two present values must be equal in equilibrium,

$$G_t/R_t = E_t(S_{t+1})/E_t(R_{t+1}^s). \qquad (5.5)$$

The CAPM framework expresses the riskiness of an asset by the following equilibrium relation,

$$E_t(R_{t+1}^s) - R_t = \beta^s[E_t(R_{t+1}^M) - R_t], \qquad (5.6)$$

where R_{t+1}^M is the one period return on the market portfolio and $\beta^s \equiv C(R_{t+1}^s; R_{t+1}^M)/V(R_{t+1}^M)$. By solving (5.5) for $E_t(R_{t+1}^s)$ and substituting the result into (5.6), one finds

$$E_t(S_{t+1})(R_t/G_t) - R_t = \beta^s[E_t(R_{t+1}^M) - R_t], \qquad (5.7)$$

which may be rewritten as

$$[E_t(S_{t+1} - G_t)/G_t]R_t = \beta^s[E_t(R_{t+1}^M) - R_t] \qquad (5.8)$$

Since G_t, R_t, and S_t are all in the time t information set, multiplication of (5.8) by G_t and division by R_t and S_t gives (5.2) with the conditions that the benchmark portfolio is the market portfolio and that the betas in (5.2) are constant. In the static CAPM the market portfolio is a legitimate benchmark while this need not be the case in the intertemporal CAPM.

The static CAPM is given empirical content by the assumption of rational expectations which allows the substitution of realizations for expectations in (5.7) and by the choice of a benchmark return. Robichek and Eaker [176] use the 30-day Eurodollar interbank deposit rate as the risk-free interest rate in the return R_t, and they measure R_{t+1}^M as the return on the Standard and Poors 500 adjusted for dividends.[35]

---

[35] Since the publication of the Roll [179] critique, there has been considerable controversy surrounding the potential for testing the static CAPM. See Gibbons [83], Stambaugh [199] and Gibbons and Ferson [84] for some recent discussion of these issues. Although the Standard and Poors 500 clearly excludes many assets from the true market portfolio, Stambaugh [199] found that such exclusions may not be that important from an empirical perspective. Shanken [193] investigates inference within a CAPM framework when one acknowledges that the benchmark portfolio is a proxy, but one is willing to specify a priori the correlation of the proxy with the true market portfolio.

TABLE X
Tests of a foreign exchange capital asset pricing model from Robichek and Eaker
[176]

Currency	Number of observations	α	β	$\bar{R}^2$	D.W.
Belgian franc	16	−0.0061393 (−1.1338)	0.34260 (3.3102)	0.40	1.975
UK pound	30	−0.0040069 (−0.90725)	0.061933 (0.85413)	−0.01	1.688
Canadian dollar	29	0.0007909 (0.50307)	0.037848 (1.4853)	0.04	0.912
Dutch guilder	28	−0.00267 (−0.47810)	0.24537 (2.7541)	0.20	1.504
French franc	29	0.0013671 (0.25654)	0.17745 (2.0606)	0.10	1.845
Italian lira	30	−0.0021093 (−0.31419)	0.00082 (0.0074398)	−0.04	1.728
Japanese yen	30	0.0000364 (0.011415)	0.10943 (2.0879)	0.10	1.479
Swiss franc	30	0.0035662 (0.56348)	0.16999 (1.6360)	0.05	1.2588
Deutsche mark	30	0.0008352 (0.12487)	0.10980 (2.1622)	0.11	1.2765

Note: The numbers in parenthesis are t-statistics. The sample period is June 1973 to June 1976 with nonoverlapping 30-day intervals.

The results of Robichek and Eaker [176] are reported in Table X which reproduces their Table 1. Notice in the case of Canada, Switzerland, and Germany that the Durbin–Watson statistic is below its lower bound which indicates the presence of first-order serial correlation and negates the interpretation of the t-statistics. This is, though, evidence against the model. For Belgium, the Netherlands, France, and Japan the t-statistics indicate statistically significant β's in (5.6) and statistically insignificant constant terms which is consistent with the theory.

In the conclusions of Robichek and Eaker [176] they note that it would be desirable to relate the β's in the structural equations to a larger model of the determination of exchange rates that specifies the different characteristics of countries that lead to different risk premiums. Such studies require the specification of more market fundamentals than just asset returns. Robichek and Eaker also recognized that a more appropriate benchmark return might be the return on a world market portfolio although difficulties in accurately measuring such a return are well-known.

5.1c. A latent variable model

Hansen and Hodrick [98] develop a way of examining the implications of (5.1) without taking a stand on the measurement of the appropriate benchmark portfolio. Their empirical specification of (5.1) assumes that β_t^G is a constant. This assumption allows development of a set of tests, but it formally removes the tests from being explicit tests of a general equilibrium model such as is described in Section 2.

The empirical model of Hansen and Hodrick [98] can be written in a compact two-equation format. The first set of equations corresponds to the equation (5.1) written for several currencies,

$$y_{t+k} = \beta x_t + u_{t+k}. \tag{5.9}$$

Here y_{t+k} is a vector of normalized forecast errors with typical element, $y_{t+k}^i \equiv (S_{t+k}^i - G_{t,k}^i)/S_t^i$, and the expected excess return on the benchmark portfolio is $x_t \equiv E_t(R_{t+1,k}^b - R_{t,k})$. The vector β corresponds to the appropriate elements from β^G, and u_{t+k} is the vector of conditional expectation forecast errors with typical element, $u_{t+k}^i = y_{t+k}^i - E_t(y_{t+k}^i)$. The vector stochastic process u_t satisfies the conditions

$$E(u_t u_{t-j}') = \begin{cases} \Omega_j & j = 0, \ldots, k-1 \\ 0 & j \geq k \end{cases} \tag{5.10}$$

$$E(u_{t+k} h_t) = 0, \tag{5.11}$$

for all h_t in the time t information set which includes x_t. Condition (5.10) indicates that, in general, the forecast errors will be contemporaneously correlated, and if $k > 1$, that is, if the forecast interval is greater than the sampling interval, the forecast errors will be serially correlated. Condition (5.11) merely reiterates the rational expectations assumption under which estimation can proceed.

Since x_t is unobservable, Hansen and Hodrick [98] recognize that testable restrictions can be derived by considering the best linear prediction of x_t based on a subset of information available at time t, as in

$$x_t = \alpha_0 + \alpha_1' z_t + \varepsilon_t. \tag{5.12}$$

Their choice of instruments for z_t is the set of past forecast errors, y_t. This is motivated by availability of data, the need to keep the number of elements in z_t small for computational purposes, and by their observation of the usefulness of y_t in predicting y_{t+k} in unconstrained environments such as their previous research, Hansen and Hodrick [97]. The error term in (5.12), ε_t, is orthogonal to y_t and has mean zero from the properties of best linear predictors. It need not be serially uncorrelated.

Substitution of y_t for z_t in (5.12) and the result into (5.11) produces

$$y_{t+k} = \beta \alpha_0 + \beta \alpha_1' y_t + v_{t+k}, \tag{5.13}$$

which is a constrained vector autoregression of y_{t+k} on a constant and y_t. The new error process is $v_{t+k} = u_{t+k} + \beta \varepsilon_t$, and by the properties of u_{t+k} and ε_t, v_{t+k} is also orthogonal to y_t. The specification of the model does not imply that v_{t+k} is orthogonal to y_{t-i} for $i \geq 1$.

Estimation of the k-step-ahead forecasting equation for y_{t+k}, given y_t, subject to the nonlinear cross-equation restrictions in (5.13) allows recovery of consistent estimates of β and $\alpha' = (\alpha_0, \alpha_1')$, once one of the elements of β is normalized to one. The normalization is required because the derivation of the relationship (5.1) is not unique. Multiplication of the random weight, ω_t, in (2.21) by an arbitrary scalar produces an alternative benchmark return that also satisfies (5.1). For discussion of the estimation technique take the normalized β_j to be β_1.

Several strategies were available for estimating the parameter vector $\delta_0 = (\beta_2, \ldots, \beta_p, \alpha')$ in (5.13). Hansen and Hodrick [98] discuss the alternative merits and computational problems inherent in various strategies. One strategy is to impose the additional assumption that

$$E(v_{t+k}/y_t, y_{t-1}, y_{t-2}, \ldots) = 0 \tag{5.14}$$

and to estimate the parameters via maximum likelihood. Typically, a Gaussian density function is employed, but it is not necessarily assumed that y_t is Gaussian. It is also often the case that some time or frequency domain approximation to the likelihood function is employed to ease the computational burden. Even with these approximations, though, Hansen and Hodrick [98] note that maximum likelihood estimation in this type of environment can be computationally intractable when the number of currencies, p, and the forecast interval, k, substantially exceed one. The problem arises because of the multivariate moving average that is induced when the forecast interval exceeds the sampling interval. Maximum likelihood estimation requires that all of the parameters of the vector moving average process be estimated simultaneously with the structural parameters of interest. An additional problem with maximum likelihood involves the potential for misspecification inherent in imposing the additional auxiliary assumption (5.14). If this assumption is false, the technique is misspecified.

A second strategy for estimation of the parameters of (5.13) is the one that is employed by Hansen and Hodrick [98]. The procedure is a version of Hansen's [96] Generalized Method of Moments (GMM). The GMM procedure allows imposition of the cross-equation restrictions implied by the latent variable model either with or without the additional assumption (5.14). Hansen and Hodrick [98] note that estimation still requires a nonlinear numerical search as in maximum likelihood, but the search is undertaken over a much smaller parameter space. A trade-off also occurs because if (5.14) is true, maximum likelihood is asymptotically more efficient.

The orthogonality conditions

The GMM estimator, as in Section 3, minimizes a criterion function that exploits the orthogonality conditions of the model. To see this, construct a family of criterion functions that employ the same set of orthogonality conditions. For the development of the argument, let $z_t' = (y_t', 1)$, and define the matrix of reduced form parameters

$$\theta(\delta_0) = \begin{vmatrix} \alpha_1' & \alpha_0 \\ \beta_2 \alpha_1' & \beta_2 \alpha_0 \\ \vdots & \vdots \\ \beta_p \alpha_1' & \beta_p \alpha_0 \end{vmatrix} \tag{5.15}$$

where δ_0 is the parameter vector with dimension ten in this case. Hansen and Hodrick [98] employ data for five countries. Hence, δ is composed of four β's and six α's. The θ matrix is referred to as the matrix of reduced form parameters because in the unconstrained system there are 30 free parameters in the θ matrix.

The vector function of the data and the parameters of the model that summarizes the orthogonality condition is defined to be

$$f(y_{t+k}, z_t, \delta) = [y_{t+k} - \theta(\delta)z_t] \otimes z_t \qquad (5.16)$$

where $\otimes$ again denotes Kronecker product. Since z_t is in the time t information set, the expectation of f evaluated at the true parameter δ_0 is zero:

$$E[f(y_{t+k}, z_t, \delta_0)] = E[v_{t+k} \otimes z_t] = 0. \qquad (5.17)$$

There are 30 orthogonality conditions in (5.17) that are available for estimation of the ten parameters in δ_0. The GMM estimator for the parameters employs the moment estimator of (5.17) for a sample of size T,

$$g_T(\delta) = T^{-1} \sum_{t=1}^{T} f(y_{t+k}, z_t, \delta). \qquad (5.18)$$

The estimate of δ_0 is found by choosing $\delta = \delta_T$ to minimize the criterion function

$$J_T(\delta) = g_T(\delta)' W_T g_T(\delta) \qquad (5.19)$$

as in (3.35) in Section 3. As before, Hansen and Hodrick [98] chose the weighting matrix W_T optimally to produce the smallest asymptotic covariance matrix for the parameters given the set of orthogonality conditions exploited by the model.

The model's test statistic

A test of the overall restrictiveness of the model can be motivated by recognizing that δ_T is a parameter vector that is chosen to set a linear combination of the orthogonality restriction, $g_T(\delta)$, equal to zero via the first order conditions for the minimization of (5.19). Thus, there are twenty linearly independent combinations of $g_T(\delta)$ that ought to be close to zero if the null hypothesis is true. Under alternative hypotheses the elements of the reduced form matrix θ

are unrestricted and can be estimated by ordinary least squares. Relaxation of the cross-equation restrictions in this way is equivalent to setting all of the sample orthogonality restrictions in (5.18) equal to zero. Therefore, the overall test of the model can be conducted by comparing the minimized value of the criterion function when the restrictions are imposed to zero, the value of the unrestricted criterion function.

Hansen [96] demonstrates formally that $TJ_T(\delta)$ is asymptotically chi-square distributed with twenty degrees of freedom in this case when W_T is chosen optimally. Estimation of the optimal weighting matrix requires estimation of the cross-spectral density of (v', z') and application of the formulas under case (v) of Hansen [96, p. 1045]. The exchange rates used by Hansen and Hodrick [98] are US dollar values of the French franc, the Japanese yen, the Swiss franc, the UK pound, and the Deutsche mark. The data set is a semiweekly sample in which Tuesday forward rates are matched with Thursday spot rates thirty days in the future and Friday forward rates are matched with Monday spot rates.[36] The sampling procedure produces an overlapping contract structure that makes the error process v_{t+k} an eighth-order moving average if ε_t is assumed to be serially uncorrelated. Consequently, exact maximum likelihood estimation would require simultaneous estimation of forty auxiliary parameters in addition to the ten structural parameters of interest.

Estimation of the latent variable model

The sample period in the analysis is 512 observations beginning on 5 February 1976 and ending on 29 December 1980. The results of the estimation are presented in Tables XI and XII which reproduce Tables 4.4 and 4.5 from Hansen and Hodrick [98].

Table XI corresponds to the unrestricted OLS estimation of the system of equations. Hansen and Hodrick [98] test whether the set of lagged forecast errors provide a significant rejection of the

---

[36] In his comment on Hansen and Hodrick [98], Hakkio [94] correctly notes that the data are not quite aligned precisely to correspond to the delivery procedure in the forward market, since Friday forward rates generally predict Tuesday spot rates. See the discussion of settlement procedures described in Section 2.

TABLE XI
Unrestricted estimation from Hansen and Hodrick [98]

$$\frac{S^i_{t+9} - G^i_{t,9}}{S^i_t} = a_i + \sum_{j=1}^{5} b_{ij}\left(\frac{S^i_t - G^i_{t-9,9}}{S^i_{t-9}}\right) + u^i_{t+9}$$

Currency	$\hat{a}_i$ (std error) Confidence	$\hat{b}_{i1}$ (std error) Confidence	$\hat{b}_{i2}$ (std error) Confidence	$\hat{b}_{i3}$ (std error) Confidence	$\hat{b}_{i4}$ (std error) Confidence	$\hat{b}_{i5}$ (std error) Confidence	$\chi^2(5)$ All b_{ij}'s $j \geq 1 = 0$ Confidence	R^2	Resid. var.
French franc	0.297 (0.320) 0.646	-0.165 (0.142) 0.754	-0.003 (0.116) 0.018	0.252 (0.151) 0.904	-0.122 (0.088) 0.835	-0.173 (0.225) 0.557	6.738 0.759	0.068	5.698
Japanese yen	0.428 (0.420) 0.691	-0.287 (0.162) 0.932	0.204 (0.148) 0.832	0.463 (0.162) 0.996	0.090 (0.126) 0.528	-0.596 (0.272) 0.972	20.303 0.999	0.165	9.626
Swiss franc	0.328 (0.414) 0.572	-0.102 (0.292) 0.272	0.067 (0.161) 0.323	0.626 (0.161) 0.999	-0.113 (0.155) 0.533	-0.853 (0.325) 0.991	16.100 0.993	0.178	10.939
UK pound	0.568 (0.371) 0.874	0.014 (0.199) 0.056	-0.055 (0.097) 0.430	0.214 (0.112) 0.944	0.190 (0.132) 0.849	-0.406 (0.182) 0.974	8.390 0.864	0.072	6.943
Deutsche mark	0328 (0.341) 0.664	-0.281 (0.167) 0.907	-0.051 (0.123) 0.320	0.417 (0.162) 0.990	-0.034 (0.108) 0.245	-0.323 (0.194) 0.904	9.462 0.908	0.124	6.394

Note: Exchange rates: US$/foreign currency; sample: 5 February 1976 to 29 December 1980; number of observations: 512

TABLE XII
The latent variable model from Hansen and Hodrick [98]

Currency	$\hat{\beta}_i$ (std error)	Reduced form coefficients							
		$\hat{\theta}_{i0}$ (std error) Confidence	$\hat{\theta}_{i1}$ (std error) Confidence	$\hat{\theta}_{i2}$ (std error) Confidence	$\hat{\theta}_{i3}$ (std error) Confidence	$\hat{\theta}_{i4}$ (std error) Confidence	$\hat{\theta}_{i5}$ (std error) Confidence	$\chi^2(5)$ All θ_{ij}'s $j \geq 1 = 0$ Confidence	R^2
French franc	0.348 (0.178)	0.084 (0.108) 0.565	-0.033 (0.062) 0.406	0.045 (0.040) 0.733	0.157 (0.099) 0.889	0.036 (0.041) 0.617	-0.249 (0.157) 0.888	2.745 0.261	0.023
Japanese yen	1.0	0.242 (0.281) 0.611	-0.095 (0.172) 0.421	0.128 (0.092) 0.836	0.452 (0.148) 0.998	0.103 (0.104) 0.677	-0.715 (0.232) 0.998	12.943 0.976	0.152
Swiss franc	1.164 (0.316)	0.282 (0.325) 0.614	-0.111 (0.200) 0.421	0.149 (0.106) 0.843	0.526 (0.157) 0.999	0.120 (0.121) 0.681	-0.832 (0.245) 0.999	17.205 0.996	0.146
UK pound	0.421 (0.242)	0.102 (0.129) 0.569	-0.040 (0.075) 0.406	0.054 (0.048) 0.742	0.190 (0.116) 0.899	0.043 (0.049) 0.622	-0.301 (0.183) 0.901	2.941 0.291	0.009
Deutsche mark	0.659 (0.229)	0.160 (0.188) 0.603	-0.063 (0.114) 0.419	0.084 (0.063) 0.818	0.298 (0.115) 0.991	0.068 (0.070) 0.667	-0.471 (0.182) 0.990	8.398 0.864	0.066

Note: Test of the constrained model $\chi^2(20) = 18.834$. See Table XI.

unbiasedness hypothesis. Consistent with earlier results, they find substantial evidence against the hypotheses of no risk premiums and constant risk premiums. These tests are the chi-square statistics in the column that is the third from the right in Table XI. Very strong results are found for the Japanese yen and the Swiss franc, in particular.

The results for the latent variable model are presented in Table XII. The value of the test statistic of the restrictions of the model is 18.834, which is below the mean of a chi-square variable with 20 degrees of freedom. Therefore, Hansen and Hodrick [98] conclude that there is little evidence in the sample against the restrictions imposed by the model. They also conclude that the latent variable model captures most of the significant variation in the deviation of expected spot rates from forward rates while recognizing that longer time series or more powerful statistical procedures might overturn the results. In retrospect, the former hedge turns out to be correct.

A reexamination of the latent variable model

Hodrick and Srivastava [106] examine the performance of the Hansen–Hodrick model using an additional 21 nonoverlapping monthly observations. The richness of the data set was abandoned because the new data were only available through a hand-copied update. The original data are therefore sampled to produce 57 comparable nonoverlapping observations. An additional change is that Hodrick and Srivastava [106] employ the five current forward premiums as instruments whereas Hansen and Hodrick [98] employ the five lagged dependent variables. The change in instrumental variables provides a more powerful test of the model.

The first result of Hodrick and Srivastava [106] is to reexamine the model with the forward premiums as instruments and using the sampled data for the period coinciding with the initial estimation of the model by Hansen and Hodrick [98]. They find the value of the chi-square statistic describing the model's overall performance to be 24.239, which has a marginal level of significance equal to 0.232. This reconfirms that the restrictions of the model cannot be rejected using the alternative instruments for this sample period.

Table XIII presents the estimation of the model with the sample period from February 1976 to September 1982. The most striking

TABLE XIII
The latent variable model from Hodrick and Srivastava [106]

	$\hat{\beta}_i$ (std error)	Reduced form coefficients							
		$\hat{\theta}_{i0}$ (std error) Confidence	$\hat{\theta}_{i1}$ (std error) Confidence	$\hat{\theta}_{i2}$ (std error) Confidence	$\hat{\theta}_{i3}$ (std error) Confidence	$\hat{\theta}_{i4}$ (std error) Confidence	$\hat{\theta}_{i5}$ (std error) Confidence	$\chi^2(5)$ All $\hat{\theta}_{ij}$'s $j \geq 1 = 0$ Confidence	R^2
French franc	1.0	13.198 (6.164) 0.968	−0.080 (0.206) 0.302	−1.244 (0.741) 0.907	−4.467 (1.945) 0.978	0.450 (0.469) 0.663	6.080 (2.719) 0.975	5.972 0.691	0.034
Japanese yen	2.059 (0.800)	27.181 (9.052) 0.997	−0.164 (0.419) 0.304	−2.562 (1.113) 0.979	−9.199 (2.614) 0.999	0.927 (0.870) 0.713	12.521 (3.567) 0.999	19.998 0.999	0.153
Swiss franc	2.607 (0.698)	34.411 (11.473) 0.997	−0.208 (0.524) 0.308	−3.243 (1.409) 0.979	−11.646 (3.066) 0.999	1.173 (1.114) 0.707	15.852 (4.138) 0.999	22.910 0.999	0.179
UK pound	1.227 (0.373)	16.188 (7.753) 0.963	−0.098 (0.245) 0.310	−1.526 (0.860) 0.924	−5.478 (2.205) 0.987	0.552 (0.579) 0.659	7.457 (2.954) 0.988	7.258 0.798	0.026
Deutsche mark	1.184 (0.222)	15.622 (7.582) 0.961	−0.094 (0.243) 0.302	−1.472 (0.869) 0.910	−5.287 (2.268) 0.980	0.533 (0.562) 0.657	7.196 (3.130) 0.979	6.222 0.715	0.024

Test of the constrained model: $\chi^2(20) = 34.497$; confidence $= 0.977$

Note: Sample period: February 1976 to September 1982; number of observations: 78.

feature of the results is the chi-square statistic testing the twenty overall restrictions of the model. Its value is now 34.497 with a marginal level of significance of 0.023. Hence, the restrictions of the model appear to be too severe during the extended sample period. Hodrick and Srivastava [106] note that if the rejection of the unbiasedness hypothesis is associated with a time varying risk premium, then the assumptions of the latent variable model are too strong. Either the β's or the α's are not constant, or some other model of risk and return other than (5.1) is necessary to describe equilibrium in the forward market.

It is interesting to compare the results of Table XIII with the estimation of the unconstrained model presented in Table XIV which reproduces Table 3 of Hodrick and Srivastava [106]. These are ordinary least squares estimates of the unconstrained reduced form coefficients. Note the differences between the two sets of estimates. In the OLS regressions the coefficients of the instrumental variables that have weak explanatory power do not always have the same algebraic sign across currencies. This is true in the case of the constant terms and the coefficients of the forward premiums of the French franc and the UK pound although in none of the cases is the set of parameters particularly precisely estimated. Also, in the case of the coefficients that do have strong explanatory power in the unconstrained model, that is the coefficients of the Swiss franc and Deutsche mark forward premiums, the rank ordering across currencies is striking but the proportionality is not of the same order of magnitude in each case. Finally, the imposition of the constraint causes a relatively severe loss in explanatory power as measured by the R^2 for the French franc, the UK pound, and the Deutsche mark.

Stability tests of reduced-form coefficients

Given the rejection of the model of risk and return postulated in this section, it is important to reiterate that the model is a statistical hypothesis and not a precisely stated theory. Ideally, one would like to test a representation of dynamic equilibrium such as that set forth by Lucas [145] and discussed in Section 2. Currently, the demands on the data to test such a model make it an exceedingly difficult

TABLE XIV
Unconstrained estimates from Hodrick and Srivastava [106]

$$(S^i_{t+1} - G^i_t)/S^i_t = a_i + \sum_{j=1}^{5} b_{ij}(G^j_t - S^j_t)/S^j_t + u^i_{t+1}$$

	$\hat{a}_i$ (std errors) Confidence	b_{i1} (std errors) Confidence	b_{i2} (std errors) Confidence	b_{i3} (std errors) Confidence	b_{i4} (std errors) Confidence	b_{i5} (std errors) Confidence	$\chi^2(6)$ All coeffs. = 0 Confidence	$\chi^2(5)$ All b_{ij}'s $j \geq 1 = 0$ Confidence	R^2	Resid. var.
French franc	-8.436 (19.880) 0.329	-1.115 (2.085) 0.407	-1.535 (1.716) 0.629	-4.993 (3.053) 0.898	-1.482 (1.057) 0.839	10.230 (4.470) 0.978	8.391 0.789	8.329 0.861	0.079	1284.68
Japanese yen	29.008 (13.238) 0.972	-0.195 (0.956) 0.161	-6.131 (1.588) 0.999	-6.953 (3.184) 0.971	1.319 (1.294) 0.692	12.363 (4.437) 0.995	36.199 0.999	36.155 0.999	0.222	1325.61
Swiss franc	27.568 (17.401) 0.886	-0.971 (1.489) 0.485	-2.659 (2.253) 0.762	-13.024 (3.676) 0.999	0.238 (1.438) 0.131	18.146 (5.775) 0.998	19.365 0.996	18.783 0.998	0.188	1755.74
UK pound	-2.302 (11.391) 0.160	0.713 (0.584) 0.778	-1.692 (1.420) 0.767	-4.933 (2.647) 0.938	-2.793 (1.243) 0.975	10.161 (3.458) 0.997	25.833 0.999	25.715 0.999	0.163	1106.15
Deutsche mark	4.857 (18.013) 0.213	-1.638 (1.702) 0.664	-2.474 (2.141) 0.752	-2.554 (3.844) 0.494	0.043 (1.470) 0.023	5.402 (6.113) 0.623	10.006 0.876	7.928 0.840	0.080	1630.13

Note: Sample period: February 1976 to September 1982; number of observations: 78.

task, although some attempts along these lines will be discussed below.

Given the rejection of the Hansen–Hodrick model, Hodrick and Srivastava [106] inquire whether the rejection can be traced to instability in the parameter estimates of the reduced form coefficients. This is an interesting question because the theoretical model that led to (5.1) only postulates the existence of a trade-off between risk and return conditional on the information set at time t. It does not impose the restriction that the conditional betas are constant, nor does it impose the stronger restriction that the conditional covariance between the return on an asset and the return on a benchmark portfolio is constant and that the conditional variance of the benchmark portfolio return is constant.

Hodrick and Srivastava [106] investigate the stability of the reduced form parameters for the following reason. Even though the restrictions of the latent variable model are rejected relative to the unconstrained model, the cross-equation constraints that characterize the theory may be valid at all points in time but with different parameters throughout time. If the structural parameters are not time invariant, this ought to be manifest in time variation of the reduced form parameters. Alternatively, if the coefficients of the reduced form are reliably stable, then some other model of risk and return other than the particular latent variable model will have to be developed to characterize the data. There are several reasons why the coefficients of the reduced form equations

$$(S_{t+1}^i - G_t^i)/S_t^i = a_i + \sum_{j=1}^{5} b_{ij}(G_t^j - S_t^j)/S_t^j + v_{t+1}^i \qquad (5.20)$$

for $i = 1, \ldots, 5$, may not be constant. If (5.20) is interpreted as a conditional expectation, the assumption is imposed that the conditional expectation is linear in the information set. If the conditional expectation is a nonlinear function of the information set, the linear projection will, in general, involve parameters that are not constant over time. Equation (5.20) may always be interpreted as a linear prediction, but testing for stability of the coefficients of this projection requires assumptions on the error term that make the projection equivalent to a conditional expectation.

It is noted in the previous section that equations such as (5.20) are unlikely to have constant coefficients in the sense that the coefficients are likely to be functions of particular government policies in the relevant countries during particular sample periods. This provides an additional reason for parameter instability.

Finally, the statistical analysis depends on asymptotic distribution theory which requires some notion of a large sample. The sample size must be sufficiently large to allow realizations from all possible parts of the parameter space with the appropriate frequency. No one knows how sensitive the analysis is to this issue, but it seems likely that small sample problems may be present if adding a year or two of data changes inference quite drastically.

A problem in investigating stability of parameters is that most traditional tests for structural change such as the Chow [30] test or the Brown, Durbin, and Evans [28] test are based on an assumption of conditonal homoscedasticity. Hodrick and Srivastava [106] perform tests of conditional homoscedasticity analogous to the one discussed above that is used by Cumby and Obstfeld [37]. They regress the squared reduced form residuals on the levels and the squares of the five forward premiums. The results of the test indicated strong evidence against conditional homoscedasticity in the case of the French franc, the Swiss franc, and the Deutsche mark.

Given this evidence Hodrick and Srivastava [106] derive a test for parameter change that can be conducted if one is willing to consider allowing sample sizes of two distinct samples to go to infinity. In estimation of the unconstrained model, as discussed above in Section 3, the GMM estimator of the parameter vector δ_0 for a sample of size T is strongly consistent and asymptotically normally distributed. That is,

$$\sqrt{Ti}(\delta_{Ti} - \delta_0) \sim N(0, \Omega_{Ti}), \qquad (5.21)$$

where Ω_{Ti} is given by (3.48) and Ti denotes the size of the ith sample. With the additional auxiliary assumption of no serial correlation and under the null hypothesis that $\delta_{T1} = \delta_{T2}$ for two distinct samples, the test statistic

$$(\delta_{T1} - \delta_{T2})'\Omega_T^{-1}(\delta_{T1} - \delta_{T2}) \qquad (5.22)$$

is asymptotically chi-square distributed with m degrees of freedom

TABLE XV
Tests for constant coefficients from Hodrick and Srivastava [106]

Currency	Test statistic	Confidence
July 1973 to January 1976 and February 1976 to December 1980		
French franc	3.725	0.286
Japanese yen	32.633	0.999
Swiss franc	14.175	0.972
UK pound	22.877	0.999
Deutsche mark	9.220	0.838
February 1976 to December 1980 and January 1981 to September 1982		
French franc	7.101	0.688
Japanese yen	4.900	0.443
Swiss franc	5.501	0.518
UK pound	10.996	0.911
Deutsche mark	10.942	0.909
February 1976 to October 1979 and November 1979 to September 1982		
French franc	17.246	0.991
Japanese yen	4.423	0.380
Swiss franc	6.859	0.665
UK pound	10.670	0.900
Deutsche mark	6.448	0.625

where m is the dimension for δ_0, δ_{T1} and δ_{T2} are estimates of δ_0 from the two subsamples, and $\Omega_T = (\Omega_{T1}/T1 + \Omega_{T2}/T2)$.

Hodrick and Srivastava perform three sets of tests, and the results are reported here in Table XV, which reproduces their Table 5. The first set of tests examines Hansen and Hodrick's [98] conjecture that the observations from the transitional years of the flexible exchange rate period from July 1973 when their data series began until the formal ratification of the Rambouillet agreement in January 1976 should be omitted from the analysis.

The ratification amended the Articles of Agreement of the International Monetary Fund to allow countries to adopt flexible exchange rates as their *de jure* system. Prior to the ratification, flexible exchange rates were only a *de facto* system, and there was considerable discussion about the possibility of a return to a system of fixed parities. In such an environment, the ratification may have been a signal of some importance in assessing possible future government policies.

The second tests examine the hypothesis that the coefficients of the unconstrained model do not differ significantly when the 21 additional observations are added to the Hansen and Hodrick [98] sample. The results from these tests provide some evidence against the null hypothesis for the UK pound and the Deutsche mark. It is interesting to note that there is no strong evidence against the null hypothesis for the two currencies that had the most explanatory power in the constrained model, namely, the Japanese yen and the Swiss franc.

The third test compares the estimated coefficients before the Carter intervention in October 1979 and the resulting change in Federal Reserve Board operating procedures with the period from November 1979 to September 1982. This appears to be a natural point at which to perform the test given the change in US policy. Somewhat surprisingly, evidence against the null hypothesis was found only in the case of the French franc and the UK pound. The yen and the Swiss franc tests again demonstrated little evidence against the hypothesis of no structural change.

Although the evidence on structural change was not particularly strong, Hodrick and Srivastava [106] do not express confidence in the validity of the tests because of the limited sample sizes and the unknown power of the test. Instead of concluding that the linear model with constant coefficients is reasonable, though, they examine a nonlinear specification that adds squared values of the forward premiums as regressors on the right-hand side of the reduced-form model. The results are presented in Table XVI which reproduces their Table 6. Notice that the tests of the significance of the squared forward premiums have very large chi-square statistics in the equation. This indicates strong potential for nonlinearity in the specification.

Nonlinearity of the conditional expectation could be responsible for evidence against time invariance and also for evidence against conditional homoscedasticity. It is also likely that the betas in the Hansen and Hodrick latent variable model would not be constant in an environment in which the conditional expectation is nonlinear.

Of course, the usual caveat must be invoked here. It is important to remember that squaring the forward premiums places even more

TABLE XVI

Investigation of nonlinear specification from Hodrick and Srivastava [106]

$$(S^i_{t+1} - G^i_t)/S^i_t = a_i + \sum_{j=1}^{5} \{b_{ij}(G^j_t - S^j_t)/S^j_t + c_{ij}[(G^j_t - S^j_t)/S^j_t]^2\} + u^i_{t+1}$$

Currency	$\chi^2(5)$ $b_{ij} = 0, j = 1, 5$ Confidence	$\chi^2(5)$ $c_{ij} = 0, j = 1, 5$ Confidence	$\chi^2(10)$ $b_{ij} = c_{ij} = 0, j = 1, 5$ Confidence
French franc	19.171 0.998	21.741 0.999	26.961 0.997
Japanese yen	9.029 0.892	9.753 0.917	44.527 0.999
Swiss franc	22.782 0.999	23.163 0.999	59.059 0.999
UK pound	8.578 0.873	10.900 0.947	68.693 0.999
Deutsche mark	13.119 0.978	13.292 0.979	23.633 0.991

Note: Sample period: February 1976 to September 1982; number of observations: 78.

weight on their large values. If statistical or small sample problems with the data are associated primarily with these values, such a procedure is probably an extremely bad way to conduct statistical inference especially when that inference must be based on large sample distribution theory. Clearly, more work needs to be done in this area before we will know which interpretation is correct, if either is.

Incorporation of other asset returns

One interesting direction to pursue the latent model is to incorporate additional returns. Campbell and Clarida [29] investigate whether the latent variable model of Hansen and Hodrick [98] can be rejected as an explanation of the predictability and comovement of risk premiums in the forward market and in the term structure of interest rates. They demonstrate that variables, such as the forward premiums, that have been useful as proxies for risk premiums in the forward foreign exchange market are also useful in predicting the excess returns on risky holding-period investments over the riskless deposit rate. Similarly, yield curve differentials from the Eurocurrency term structure also are demonstrated to have power in the forward market. Campbell and Clarida [29] are unable to reject the hypothesis that the expected dollar returns on uncovered three-month Eurodeutschemark and Eurosterling deposits and those on rolling over one-month deposits of these currencies versus holding a three-month deposit in the currency move in proportion to a single latent variable. Their study utilizes six years of weekly data with a three-month overlap.

Giovannini and Jorion [86] have also extended the latent variable model to allow time variation in the betas. They use weekly data on the excess returns of the UK pound, the Deutsche mark, the Dutch guilder, and the Swiss franc relative to the US dollar, as well as the excess return of the US stock market relative to the Eurodollar rate. The specification allows the betas in equations like (5.1) to be deterministic functions of the nominal interest rates of the US and the other country. The instruments include a constant and the five nominal interest rates.

For a sample from 12 July 1974 to 28 December 1984 (547 observations), the chi-square statistic that tests the overidentifying restriction of the model has a marginal level of significance of 0.339. Furthermore, the hypothesis that the beta coefficients of the nominal interest rates are all zero is grossly rejected by the data.

5.1d. An ARCH model of the risk premium

An alternative model of the risk premium that also does not utilize additional data other than exchange rates is investigated by Domowitz and Hakkio [42]. This paper is an interesting mix of current theory and new econometric analysis. The theory is taken from Lucas [145], and the econometric analysis builds on the work of Engle [55]. The goal of their study is to model a time varying risk premium using the autoregressive conditional heteroscedasticity (ARCH) framework.

As noted above in Section 4.1, Domowitz and Hakkio [42] assume that the exogenous processes of the Lucas [145] model are conditionally log-normal as in (5.23a–d). They further constrain the time series processes to be first order autoregressive as in

$$x_t = \rho_1 x_{t-1} + u_{1t}, \qquad (5.23a)$$

$$y_t = \rho_2 y_{t-1} + u_{2t}, \qquad (5.23b)$$

$$m_t = \gamma_1 m_{t-1} + u_{3t}, \qquad (5.23c)$$

and

$$n_t = \gamma_2 n_{t-1} + u_{4t}. \qquad (5.23d)$$

Here the lower case letters (x, y, m, n) are natural logarithms of the endowments of the two countries and their two monies, and the vector of innovations $u_t' = (u_{1t}, u_{2t}, u_{3t}, u_{4t})$ is assumed to be a conditional Gaussian process with mean zero and condition covariance matrix $H_t = \mathrm{diag}(h_{1t}, h_{2t}, h_{3t}, h_{4t})$. All off-diagonal elements of the covariance matrix are assumed to be zero.

Domowitz and Hakkio [42] also assume that preferences in (2.1) are given by a Cobb–Douglas form,

$$U(X_t, Y_t) = A X_t^\alpha Y_t^{(1-\alpha)}, \qquad (5.25)$$

in which case it can be demonstrated that (2.5) reduces to

$$S_t = [(1 - \alpha)/\alpha](M_t/N_t), \qquad (5.26)$$

and the one period intertemporal marginal rates of substitution of monies in (2.7) and (2.10) are

$$Q_{t+1}^m = \beta[(Y_{t+1}/X_{t+1})^{1-\alpha}/(M_{t+1}/X_{t+1})]/[(Y_t/X_t)^{1-\alpha}/(M_t/X_t)] \qquad (5.27a)$$

and

$$Q_{t+1}^n = \beta[(X_{t+1}/Y_{t+1})^\alpha/(N_{t+1}/Y_{t+1})][(X_t/Y_t)^\alpha/(N_t/Y_t)]. \qquad (5.27b)$$

From the expression for interest rate parity in (2.11) the forward rate is

$$G_t = S_t E_t(Q_{t+1}^n)/E_t(Q_{t+1}^m). \qquad (5.28)$$

Equations (5.23) imply the following expressions for $E_t(s_{t+1})$ and g_t:

$$E_t(s_{t+1}) = \ln[(1 - \alpha)/\alpha] + \gamma_1 m_t - \gamma_2 n_t \qquad (5.29)$$

and

$$g_t = \ln[(1 - \alpha)/\alpha] + \gamma_1 m_t - \gamma_2 n_t - (1/2)(h_{3t+1} - h_{4t+1}).^{37} \quad (5.30)$$

Consequently, the logarithmic representation of the risk premium is

$$E_t(s_{t+1}) - g_t = (1/2)(h_{3t+1} - h_{4t+1}). \quad (5.31)$$

In this simplified framework, the risk premium depends only on the difference between the conditional variances of the two money supplies. An increase in the conditional variance of the home money, h_{3t+1}, increases the risk premium. The result occurs because there is no effect on the expected logarithm of the future spot rate while the logarithm of the forward rate falls with the decrease in domestic interest rates. Domestic interest rates fall because an increase in the variance of the domestic money increases the variance of the purchasing power of the money which contributes positively to the return on nominal assets denominated in that money. Expected returns on other assets must rise to compensate investors in equilibrium.

From (5.27) the rational expectations forecast error of the logarithm of the spot rate is $\varepsilon_{t+1} \equiv s_{t+1} - E_t(s_{t+1}) = u_{3t+1} - u_{4t+1}$, and (5.31) can be rewritten as

$$s_{t+1} - s_t = p_t + (g_t - s_t) + \varepsilon_{t+1}, \quad (5.32)$$

where the risk premium is $p_t \equiv (1/2)(h_{3t+1} - h_{4t+1})$, while the conditional variance of ε_{t+1} is $(h_{3t+1} + h_{4t+1})$. If h_{3t+1} and h_{4t+1} vary through time, there will be a time varying risk premium and the error term in (5.32) will exhibit conditional heteroscedasticity.

In their econometric analysis Domowitz and Hakkio [42] do not impose the restrictions derived above. Instead, they use them as a motivation for investigation of a simpler, single equation model that works directly with a version of (5.32) and assumes that the conditional variance of ε_{t+1} is generated from an ARCH process. They also assume that the risk premium, p_t, depends directly on the

---

[37] The expression in (5.30) requires expressions for $\ln[E_t(Q_{t+1}^m)]$ and $\ln[E_t(Q_{t+1}^n)]$. From the log-normal distributions (5.23) it can be demonstrated that

$$E_t(Q_{t+1}^m) = \exp\{\ln \beta - \alpha(1 - \rho_1)x_t - (1 - \alpha)(1 - \rho_2)y_t$$
$$+ (1 - \gamma_1)m_t + (\alpha^2/2)h_{1t+1} + [(1 - \alpha)^2/2]h_{2t+1} + (1/2)h_{3t+1}\}.$$

A similar expression can be derived for $E_t(Q_{t+1}^n)$.

conditional variance of ε_{t+1} that is denoted h_{t+1}^2. Notice that the theory does not predict this specification. The exact specification of their model is

$$(S_{t+1} - S_t)/S_t = P_t + \beta_1(G_t - S_t)/S_t + \varepsilon_{t+1} \qquad (5.33a)$$

$$P_t = \beta_0 + \theta h_{t+1} \qquad (5.33b)$$

$$\varepsilon_{t+1} \mid I_t \sim N(0, h_{t+1}^2) \qquad (5.33c)$$

$$h_{t+1}^2 = \alpha_0^2 + \sum_{j=1}^{J} \alpha_j^2 \varepsilon_{t+1-j}^2 \qquad (5.33d)$$

where I_t represents the informational set at time t.[38] The conditional variance of the rate of depreciation given time t information, is postulated to depend on the realizations of the J previous squared error terms. Domowitz and Hakkio [42] set $J = 4$ in their empirical analysis, and since the conditional variance enters the regression model, they refer to their model as an ARCH-in-mean (AIM) model. This type of model was originally introduced into the literature by Engle, Lilien and Robins [56].

The likelihood function and estimation

Estimation of the AIM model requires maximum likelihood techniques. The log-likelihood function for T observations is

$$L = T^{-1} \sum_{t=1}^{T} L_t \qquad (5.34a)$$

$$L_t = -\log h_{t+1} - (1/2)\varepsilon_{t+1}^2/h_{t+1}^2 \qquad (5.34b)$$

with the constant term suppressed. This likelihood can be maximized with respect to the unknown parameters $\delta_0 = (\beta_0, \beta_1, \theta, \alpha_0, \alpha_1, \ldots, \alpha_J)$. The estimation strategy is straightforward, but the derivatives of the likelihood function are relatively

---

[38] Domowitz and Hakkio [42] use the same data as Hodrick and Srivastava [106], but, in constrast, they use all 108 observations in one specification. They conduct a number of specification tests that are discussed below. They note that the correlations between the logarithmic representations of the rates of depreciation and the forward premiums that are derived in theory and the specification they use in the empirical analysis exceed 0.999 in all cases except the French franc rate of depreciation for which the correlation is 0.988.

complicated expressions because of the presence of h_{t+1} in the mean.

As with other maximum likelihood techniques, it can be shown, under suitable regularity conditions on the data, that

$$\sqrt{T}(\delta_T - \delta_0) \xrightarrow{D} N(0, I_\delta^{-1}),$$

where δ_T is an estimate based on T observations, and I_δ is the complete information matrix.

The results of the estimation are presented in Table XVII which reproduces Table 2 of Domowitz and Hakkio [42]. Two specification tests are presented that are designed to check the adequacy of the model prior to tests of economic hypotheses.

The first test is a general heteroscedasticity test. Under the null hypothesis, there should be no residual heteroscedasticity, and a residual-based diagnostic test of this is conducted by regressing $(\hat{\varepsilon}_{t+1}^2 - \hat{h}_{t+1}^2)/\hat{h}_{t+1}^2$ on $1/\hat{h}_{t+1}^2$, $x_t' x_t / \hat{h}_{t+1}^2$ and $x_t' \hat{h}_{t+1}/\hat{h}_{t+1}^2$ where $x_t \equiv (G_t - S_t)/S_t$, and by testing whether the coefficients on these variables are statistically significantly different from zero. The results of the test, labelled het in Table XVII, indicate some evidence against the hypothesis for the Japanese yen and the Deutsche mark because the marginal levels of significance are 0.036 and 0.074, respectively.[39]

The second specification test corresponds to a version of the information matrix test proposed by White [215]. This is a general specification test of the validity of the model against any alternative that renders invalid the usual maximum likelihood inference. This White procedure tests the equivalence of two ways of expressing the information matrix, in Hessian form, $-E[\partial^2 L/\partial\delta\partial\delta']$, or in outer product form, $E[\partial L/\partial\delta \cdot \partial L/\partial\delta']$. White [215] notes that, under appropriate regularity conditions, the "indicators" given by $\partial L_t/\partial\delta_i \cdot \partial L_t/\partial\delta_j + \partial^2 L_t/\partial\delta_i\partial\delta_j$, for $i = 1, \ldots, m$ where m is the total number of parameters, can be used to test the equivalence of the two expressions based on the asymptotic normality of the indicators.

Domowitz and Hakkio [42] perform the information matrix test

---

[39] The test could not be conducted for the Swiss franc because the estimates of α are so small that θ is essentially unidentified.

on a subset of the indicators due to insufficient degrees of freedom. The results are similar to the heteroscedasticity test except for the Japanese yen for which the information matrix test provides no evidence of misspecification of the model. Given these findings, Domowitz and Hakkio [42] conclude that the model given by (5.33) is not obviously rejected by the data. This motivates an examination of several economic hypotheses regarding the nature of risk premiums in the forward market.

The first test, done under the maintained hypothesis that $\beta_1 = 1$, tests whether $\beta_0 = 0$ and $\theta = 0$. The results labeled RP1 in Table XIV indicate little evidence against this hypothesis. If this were so, then the unbiasedness hypothesis would be correct. The second test examines the no-risk-premium hypothesis by testing $\beta_0 = 0$, $\beta_1 = 1$, and $\theta = 0$. The tests are reported in Table XVII in the row labeled RP2. This hypothesis is rejected for the UK pound and the Japanese yen.

Domowitz and Hakkio [42, p. 62] conclude that their results are generally consistent with the rejection of the unbiasedness hypothesis, but 'there is little support for the conditional variance of the exchange rate forecast error being an important sole determinant of the risk premium.' This latter statement should be qualified by noting that the conditional variance is modeled in a particularly strong specification with the ARCH process.

The ARCH process forces the conditional variance to take its largest values after the largest residual errors of the sample. It may be that such large errors occur because of the resolution of uncertainty, and the conditional variance actually is smaller after some large errors. Such a situation would presumably occur after the announcement of a change in monetary or fiscal policy that is preceded by a period of debate about the direction of the policies. Hence, the ARCH model may be a particularly bad way to model conditional heteroscedasticity in a rational world.

5.1e. Expected real interest rates and risk premiums

Another approach to developing a model of the risk premium that uses only real asset returns is developed in Korajczyk [122]. This section examines the empirical model of Korajczyk [122] that relates the risk premium in the forward foreign exchange market to

TABLE XVII
ARCH-in-mean model from Domowitz and Hakkio [42]

$$(S_{t+1} - S_t)/S_t = \beta_0 + \beta_1[(G_t - S_t)/S_t] + \theta h_{t+1} + \varepsilon_{t+1}$$
$$h_{t+1}^2 = \alpha_0^2 + \alpha_1^2 \varepsilon_{t+1-1}^2 + \ldots + \alpha_4^2 \varepsilon_{t+1-4}^2$$

	UK pound	French franc	Deutsche mark	Japanese yen	Swiss franc
β_0	-0.016 (0.002)	0.020 (0.022)	0.040 (0.023)	-0.020 (0.21)	0.010 (0.038)
β_1	-1.755 (0.884)	0.381 (0.632)	0.054 (1.120)	-0.211 (0.444)	-1.092 (1.436)
θ	0.325 (0.665)	-0.708 (0.798)	-1.120 (0.714)	0.723 (0.806)	0.006 (0.010)
α_0	0.023 (0.003)	0.024 (0.003)	0.027 (0.004)	0.021 (0.004)	0.036 (0.003)
α_1	0.555 (0.207)	0.378 (0.176)	0.209 (0.213)	0.126 (0.480)	$0.211*10^{-6}$ ($0.211*10^{-6}$)
α_2	$0.447*10^{-7}$ (0.190)	$-0.109*10^{-7}$ (0.189)	$0.242*10^{-7}$ (0.379)	0.147 (0.225)	$0.162*10^{-9}$ ($0.162*10^{-9}$)
α_3	$0.103*10^{-7}$ (0.365)	0.458 (0.171)	0.476 (0.158)	0.485 (0.144)	$0.125*10^{-7}$ ($0.457*10^{-7}$)
α_4	0.100 (0.330)	$0.505*10^{-7}$ (0.523)	0.501 (0.184)	0.495 (0.160)	$0.636*10^{-7}$ ($0.636*10^{-7}$)
RP1	0.125	0.681	0.159	0.573	0.500
RP2	0.024	0.612	0.275	0.017	0.518
het	0.358	0.271	0.074	0.036	—
Info	0.742	0.485	0.107	0.810	0.742

Notes: Standard errors are in parentheses below each coefficient. The number reported for RP1, RP2, het, and Info are marginal significance levels. RP1 is an F test of the hypothesis $\beta_0 = 0$ and $\theta = 0$: RP2 is an F test of the hypothesis $\beta_0 = 0$, $\beta_1 = 1$ and $\theta = 0$; het is the residual based test of remaining heteroscedasticity; Info in the information matrix test of White [215] for general model specification.

the deviation between the two countries' expected real interest rates. Mishkin [164] strongly rejects the hypothesis that the US real interest rate is constant, and the hypothesis of expected real rate equality has been rejected by Cumby and Obstfeld [37] and Mark [149].

Korajczyk's [122] empirical model is constructed from three building blocks. The first is interest rate parity as given in (2.11). The second is a definition of the deviation from absolute purchasing power parity at time t that is denoted D_t, where

$$D_t \equiv kP_t^* S_t / P_t, \tag{5.35}$$

where k is a constant. If purchasing power parity held, the exchange rate would be equal to the ratio of the domestic price level, P_t, to the foreign price level, P_t^*, and D_t would be constant.

By combining (2.11) and (5.35) and taking logarithms, Korajczyk [122] derives an expression for the ex post difference between the logarithm of the future spot rate, s_{t+1}, and the logarithm of the current forward rate. The expression is

$$s_{t+1} - g_t = \Delta p_{t+1} - \Delta p_{t+1}^* + \Delta d_{t+1} - (R_t - R_t^*) \tag{5.36}$$

where Δ is the difference operator and R_t and R_t^* are the risk-free one-period nominal returns on bills denominated in the domestic and foreign currencies, respectively.

The third building block is a Fisher equation for each country that expresses the nominal risk-free return as the sum of the expected rate of inflation and the expected real return:

$$R_t = E_t(r_{t+1}) + E_t(\Delta p_{t+1})$$
$$R_t^* = E_t(r_{t+1}^*) + E_t(\Delta p_{t+1}^*). \tag{5.37}$$

Substitution of (5.37) into (5.36) and projection of both sides onto the time t information set gives

$$E_t(s_{t+1}) - g_t = E_t(r_{t+1} - r_{t+1}^*) + E_t(\Delta d_{t+1}). \tag{5.38}$$

Korajczyk [122] assumes that logarithmic deviations from purchasing power parity follow a random walk in which case $E_t(\Delta d_{t+1}) = 0$. This assumption is motivated by appeal to Roll's [180] "efficient market version of PPP," and by the empirical work of Adler and Lehmann [3] that is supportive of the hypothesis.

Korajczyk [122] notes correctly that the existence of international commodity arbitrage would lead one to expect some mean reversion in the d_t series in contrast to the random walk hypothesis.[40] Cumby and Obstfeld [37] report evidence against this hypothesis since they demonstrate that the expected inflation differential is not equal to the expected rate of change of the exchange rate.

Korajczyk [122] derives an estimable specification by taking the projection of both sides of (5.38) onto an observable set of time t information, X_t,

$$E[(s_{t+1} - g_t) \mid X_t] = E[(r_{t+1} - r_{t+1}^*) \mid X_t]. \tag{5.39}$$

Then, using the assumption of rational expectations, he derives

$$s_{t+1} - g_t = \theta_0 + \theta_1 E[(r_{t+1} - r_{t+1}^*) \mid X_t] + \theta_2 Z_t + \varepsilon_{t+1} \tag{5.40}$$

where Z_t is any subset of X_t and ε_{t+1} is orthogonal to X_t. The theory implies that θ_0 and θ_2 should equal zero and θ_1 should equal unity. Under the null hypothesis the residual ε_{t+1} is uncorrelated with X_t, but it is not necessarily serially uncorrelated.

Korajczyk [122] imposes the additional assumption that ε_{t+1} is serially uncorrelated, and he estimates several specifications like (5.40) with three stage least squares (3SLS). This technique is employed because some first stage regressions are necessary to obtain $E[(r_{t+1} - r_{t+1}^*) \mid X_t]$. The variables in the first stage regressions are a constant, the average realized real interest rate differential over the previous twelve months, the inflation differential between the United States and the other country in the previous month, the difference across countries in the sample standard deviation in nominal interest rates over the previous 26 weeks, the lagged dependent variable, and the forward premium, $g_t - s_t$, at the beginning of the month.[41]

The sample period for the analysis is monthly observations from

---

[40] Samuelson [189] argues that although not much empirical evidence may exist against the hypothesis that logarithms of real stock prices follow a random walk, economic law suggests that a firm is unlikely to become infinitely valuable which is allowed by that stochastic process. Similarly, it seems difficult to conceive that the real value of one currency relative to another is likely to go to either zero or infinity. Hence, although analysis of some data sets seems to support Korajczyk's [122] hypothesis, economic theory suggests it is probably false.
[41] Korajczyk [122] explains the choice of instrumental variables by reference to their predictive power in previous financial studies.

April 1974 to December 1980, a total of 81 observations. The exchange rate data are US dollar values of the UK pound, the Canadian dollar, the French franc, the Swiss franc, the Italian lira, the Deutsche mark, the Dutch guilder, and the Belgian franc. Consumer price indexes are taken from *International Financial Statistics* to construct the inflation rates, and Eurocurrency interest rates are used for risk-free nominal returns.

The first tests that Korajczyk [122] performs are unconditional tests of (5.56). The Hotelling T^2 statistics test whether the means of the forward rate forecast errors, $(s_{t+1} - g_t)$, the realized real interest rate differentials, $(r_{t+1} - r_{t+1}^*)$, and the differences of these variables, $(s_{t+1} - g_t) - (r_{t+1} - r_{t+1}^*)$, are significantly different from zero. The joint hypothesis that all the means are equal to zero is rejected at the 0.05 marginal level of significance for the set of forward rate forecast errors and the realized real rate differentials, but the hypothesis cannot be rejected for the differences between the two variables. While there may be some reason to doubt the validity of the test statistics due to nonnormality of the realizations or conditional heteroscedasticity, the results are certainly suggestive that risk premiums in the forward market are related to real interest differentials.

In his 3SLS analysis of (5.40) Korajczyk [122] first performs the estimation without any Z_t variables as in

$$s_{t+1} - g_t = \delta_0 + \delta_1 E_t[(r_{t+1} - r_{t+1}^*) \mid X_t] + \varepsilon_{t+1}. \tag{5.41}$$

In conducting tests of hypotheses Korajczyk [122] examines three alternative ways of generating distributions for the test statistics. They are the asymptotic distribution from 3SLS, a bootstrap distribution, and a Monte Carlo simulation.

When error distributions have fatter tails than if they are normally distributed, convergence to the asymptotic distribution may be slower than is implied by the theory. The asymptotic distribution theory is correct in large samples as long as the fat-tailed distribution has a finite variance (e.g., student-*t* rather than Pareto-Levy distributions). Korajczyk's [122] findings indicate that there may be bias in the asymptotic distribution theory because the results of the bootstrap and Monte Carlo simulations suggest somewhat different inference.

The bootstrap (see Efron [52]) is a nonparametric procedure for producing small-sample properties of test statistics. In this case errors from a vector of regression equations are assumed to be drawn from some unspecified multivariate distribution function denoted F. An estimate for this joint distribution is provided by the residuals from the original regressions, the "empirical" error distribution that is denoted $\hat{F}$. This distribution is constructed by placing probability mass $1/T$ on $\hat{\varepsilon}_t$, where $\hat{\varepsilon}_t$ is the vector of residuals at time t, $t = 1, \ldots, T$. A small sample distribution of the test statistics under the null hypothesis is generated by (i) sampling from the empirical error distribution with replacement, (ii) constructing regression equations such that the null hypothesis is true, and (iii) reestimating the regression and calculating the desired statistics. Repeatedly following these steps leads to a distribution for each test statistic from which percentiles can be calculated.

Korajczyk [122] notes several potential problems concerning the bootstrap distribution. Much of the uncertainty about the quality of bootstrap estimates involves how well the empirical error distribution, $\hat{F}$, approximates the true distribution, F. In a least-squares regression, outliers can significantly affect regression coefficients, and, hence, the estimated regression errors that form the distribution $\hat{F}$ thereby making $\hat{F}$ relatively far from F. Korajczyk [122] argues that this may be a particularly troublesome problem in financial data which tend to have fat tails and consequently a greater probability of such outliers.

A second problem concerning the validity of bootstrap distributions involves the falsely optimistic results that can arise if significant overfitting or pretesting is done. In this case $\hat{F}$ will understate F.

A third criticism that is applicable to both the bootstrap and the Monte Carlo simulations that is not mentioned by Korajczyk [122] involves his treatment of the right-hand-side variables. Bootstrap and Monte Carlo techniques typically employ assumptions that the right-hand-side variables are fixed in repeated samples or are strictly exogenous. If these variables are merely predetermined endogenous variables, as they are in most of the analyses discussed in this monograph, such techniques may induce an error because they do not generate new right-hand-side variables whose values are, in theory, simultaneously determined with the realizations of the new errors. The degree to which this problem invalidates the simulation results is, of course, unknown.

The Monte Carlo simulations are also potentially useful for

assessing the influence of small sample size. The idea here is to assume that the covariance matrix of the residuals is the true covariance matrix of a multivariate normal distribution and to draw errors from a random number generator using that distribution. Differences between the Monte Carlo distribution and the asymptotic distribution of the test statistics are therefore due to small sample size (within sampling error).

The results of Korajczyk's [122] unrestricted estimation of the system of equations (5.41) are presented in Table XVIII which reproduces Table 3 of his paper. Note that use of the asymptotic distribution of the test statistics and 5 percent as the size of the test results in a failure to reject the null hypothesis $\delta_0 = 0$ for all currencies, but it provides a rejection of the three other joint hypotheses (equality of δ_1 across currencies, $\delta_1 = 1$ for all currencies, and $\delta_0 = 0$ and $\delta_1 = 1$ for all currencies). In contrast the bootstrap and Monte Carlo estimates of the distribution of the test statistics indicate less evidence against these hypotheses because they tend to have higher probabilities of much larger values than the asymptotic distribution. Korajczyk's [122] Figures 1–4 are particularly striking evidence of these phenomena.

Given these findings Korajczyk [122] constrains the δ_1 coefficients to be equal across currencies which results in an estimated value of δ_1 equal to 1.37 with an asymptotic standard error of 0.29, although he correctly notes that an appropriate standard error ought to take into consideration the pretest that occurred.

The second set of tests performed by Korajczyk [122] includes the lagged dependent variable as the Z_t in (5.40). This specification is employed because autocorrelation in the forward rate forecast errors as in Hansen and Hodrick [97, 98] is a source of rejection of the unbiasedness hypothesis. The results of the estimation and tests are presented in Table XIX that reproduces Korajczyk's Table 5.

The tests indicate that some evidence exists against the joint hypothesis that $\theta_0 = 0$, $\theta_1 = 1$, and $\theta_2 = 0$ for all currencies. The marginal significance levels of the Monte Carlo and bootstrap distributions of 0.05 and 0.04 are, though, somewhat different from the 0.003 of the asymptotic distribution. Korajczyk [122] concludes therefore that the highly significant autocorrelation of forward rate forecast errors appears to be a result of time variation in expected real interest differentials.

TABLE XVIII
Unrestricted instrumental variables estimates from Korajczyk [122]

$$\bar{s}_{t+1} - g_t = \delta_0 + \delta_1(\bar{r}^*_{t+1} - \bar{r}_{t+1}) + \varepsilon_{t+1}$$

A. Estimates

Currency	δ_0	δ_1	λ for $\delta_0 = 0,\ \delta_1 = 1$	D-W[a]
UK pound	0.010 (0.005)	3.10 (1.51)	5.38	1.85
Canadian dollar	−0.002 (0.001)	1.78 (1.02)	3.04	1.94
Belgian franc	0.003 (0.004)	0.64 (0.48)	1.31	1.99
French franc	0.001 (0.004)	2.80 (1.14)	3.98	2.37
Deutsche mark	0.001 (0.005)	0.21 (1.77)	0.40	1.98
Italian lira	−0.001 (0.003)	2.32** (0.57)	9.41**	2.28
Dutch guilder	0.000 (0.004)	0.74 (0.47)	0.48	2.19
Swiss franc	0.001 (0.005)	0.84 (1.56)	0.15	2.04

B. Test statistics

Test	λ	df	P-value asymptotic	Monte Carlo	P-value Bootstrap
$\delta_0 = 0$: all equations	15.06	8	0.058	0.084	0.137
δ_1: equal across equations	15.54	7	0.030	0.007	0.090
$\delta_1 = 1$: all equations	18.00	8	0.021	0.057	0.083
$\delta_0 = 0$ and $\delta_1 = 1$: all equations	28.63	16	0.027	0.097	0.130

Note: Asymptotic standard errors in parentheses; 3SLS estimates; weighted $R^2 = 0.07$.

[a] D-W statistic from first-stage regressions.

* Significantly different from the null at the 0.05 level, using asymptotic standard errors.

** Significantly different from the null at the 0.01 level, using asymptotic standard errors.

The test statistic, λ, for the linear restriction $r = R\theta$ (where θ is the parameter vector) is given by:

$$K\frac{(r - R\hat{\theta})'[R \operatorname{cov}(\hat{\theta})R']^{-1}(r - R\hat{\theta})}{\hat{\varepsilon}' \operatorname{cov}(\hat{\varepsilon})^{-1}\hat{\varepsilon}},$$

where $\operatorname{cov}(\hat{\theta}) =$ the estimate of the covariance matrix of $\hat{\theta}$; $\hat{\varepsilon} =$ the vector of regression residuals; $\operatorname{cov}(\hat{\varepsilon}) =$ the estimate of the covariance matrix of ε; and $K =$ the number of degrees of freedom. The asymptotic distribution of λ is $\chi^2(q)$, where q is the number of test restrictions (see Thiel [208], pp. 402, 508–13).

TABLE XIX
Restricted instrumental variables estimates from Korajczyk [122]

$$s_{t+1} - g_t = \theta_0 + \theta_1(r_{t+1}^* - r_{t+1}) + \theta_2(s_t - g_{t-1}) + \varepsilon_{t+1}$$

A. Estimates

Currency	$\hat{\theta}_0$	$\hat{\theta}_1$	$\hat{\theta}_2$	D-W
UK pound	0.006	1.57	0.22	2.00
	(0.004)	(0.37)	(0.11)	
Canadian dollar	−0.002	1.57	0.10	2.02
	(0.001)	(0.37)	(0.12)	
Belgian franc	0.000	1.57	0.13	1.90
	(0.004)	(0.37)	(0.07)	
French franc	0.001	1.57	0.01	2.07
	(0.004)	(0.37)	(0.10)	
Deutsche mark	−0.002	1.57	0.11	1.86
	(0.004)	(0.37)	(0.10)	
Italian lira	0.000	1.57	0.01	1.96
	(0.004)	(0.37)	(0.09)	
Dutch guilder	0.000	1.57	0.04	1.86
	(0.004)	(0.37)	(0.09)	
Swiss franc	0.002	1.57	−0.04	1.94
	(0.005)	(0.37)	(0.08)	

B. Test statistics

Test	λ	df	P-value asymptotic	P-value Monte Carlo	P-value bootstrap
$\theta_0 = 0$, all equations	10.83	8	0.212	0.350	0.353
θ_1, equal across equations	13.72	7	0.056	0.153	0.170
$\theta_1 = 1$, all equations	17.86	8	0.022	0.083	0.130
$\theta_2 = 0$, all equations	20.27	8	0.009	0.053	0.047
θ_0, $\theta_2 = 0$, and					
$\theta_1 = 1$, all equations	47.72	24	0.003	0.050	0.040

Note: Weighted $R^2 = 0.08$. See also Table XVIII.

One interesting extension of this analysis would be to include the forward premium as the variable Z_t in (5.40). Given the results of Fama [59], such a specification would be a powerful test that the risk premium in the forward foreign exchange market is only related to the expected real interest rate differential.

5.2. Models with market fundamentals

This section considers a number of recent models of the risk premium in the forward foreign exchange market. Unlike the models of the previous section, these models employ some measurements of market fundamentals (things other than exchange rates and asset returns) in an attempt to test specific theories of the risk premium.

5.2a. Models based on mean variance optimization

The first model is Frankel's [67] asset market equilibrium model based on asset demands derived from a two-period mean-variance maximization problem. The important market fundamentals are measurements of actual asset supplies. Building on the partial equilibrium model in Dornbusch [46, 47], Frankel [67] postulates that investors choose their portfolios of nominal interest earning assets that are denominated in different currencies so as to maximize a function that is increasing in expected end-of-period real wealth and decreasing in the variance of end-of-period real wealth.

Let x_t denote the vector of portfolio shares in Deutsche marks, UK pounds, Japanese yen, French francs, and Canadian dollars, with the residual share invested in US dollars, and let r_{t+1} denote the vector of real rates of return on the five currencies with $r^{\$}_{t+1}$ being the real rate of return on the US dollar. If W_t is real wealth, then Frankel [67] writes

$$W_{t+1} = W_t[x_t'(r_{t+1} - r^{\$}_{t+1}\underline{1}) + (1 + r^{\$}_{t+1})] \qquad (5.42)$$

where $\underline{1}$ is a vector of ones.

Notice that (5.42) ignores several potentially relevant economic factors. Investors will generally include other assets in their portfolios; they will receive labor or endowment income; and they will consume various real goods. Each of these factors can lead to a misspecification of the model as is noted below.

Frankel [67] employs a standard approximation for the real return on the jth country's asset given by

$$(1 + r^{j}_{t+1}) - \frac{S^{j}_{t+1}(1 + i^{j}_t)P^{\$}_t}{S^{j}_t P^{\$}_{t+1}} \doteq (1 + i^{j}_t) - \Delta p^{\$}_{t+1} + \Delta s^{j}_{t+1}, \qquad (5.43)$$

where i_t^j is the nominal interest rate on the jth country's asset, S_t^j is dollars per currency j, $P_t^\$$ is the dollar price level, lower case letters (other than nominal interest rates) are natural logarithms of their respective upper case counterparts, and Δ is the first difference operator. Therefore, the US rate of inflation is $\Delta p_{t+1}^\$$, and the rate of depreciation of the dollar relative to currency j is Δs_{t+1}^j. The real return on the US dollar asset is similarly approximated with

$$(1 + r_{t+1}^\$) \simeq (1 + i_t^\$) - \Delta p_{t+1}^\$. \tag{5.44}$$

Substituting (5.44) and (5.43) into (5.42) gives

$$W_{t+1} = W_t[x_t'(i_t - i_t^\$\underline{1} + \Delta s_{t+1}) + (1 + i_t^\$) - \Delta p_{t+1}^\$], \tag{5.45}$$

where i_t is the vector of nominal interest rates on non-dollar assets, and Δs_{t+1} is the vector of rates of depreciation of the dollar relative to these currencies. Frankel [67] assumes that the dollar price level is a geometric weighted average of the prices of goods consumed from the different countries. If α is the vector of consumption shares and $\Delta \bar{p}_t^j$ is the rate of change in the nominal price of the jth country's goods, with vector representation $\Delta \bar{p}_t$, then

$$\Delta p_t^\$ = \alpha'(\Delta \bar{p}_t + \Delta s_t) + (1 - \alpha'\underline{1})\Delta \bar{p}_t^{us}, \tag{5.46}$$

where $\Delta \bar{p}_t^{us}$ is the rate of change in the dollar price of the US good. The bars above variables represent an important assumption that Frankel [67] makes at this point: prices of goods expressed in the currency of the producing country are nonstochastic. Consequently, the only source of purchasing power risk in the evolution of the dollar price level is exchange rate uncertainty.[42]

By substituting (5.46) into (5.45) and by assuming that investors maximize

$$F[E_t(W_{t+1}), V_t(W_{t+1})] \tag{5.47}$$

with respect to x_t, the optimal portfolio shares are found to be

$$x_t = \alpha + (\rho\Omega)^{-1}[i_t - i_t^\$\underline{1} + E_t(\Delta s_{t+1})] \tag{5.48}$$

where $\rho \equiv -F_2 W_t^2/F_1$, which is the coefficient of relative risk

---

[42] Frankel [67] notes that Krugman [126] makes a similar assumption, and he argues that the assumption of sticky prices with monthly data is not particularly strong. The assumption is relaxed in Frankel and Engel [69] discussed below.

aversion, and $\Omega \equiv E_t[\Delta s_{t+1} - E_t(\Delta s_{t+1})][\Delta s_{t+1} - E_t(\Delta s_{t+1})]'$, which is the conditional covariance matrix of the relative rates of currency depreciation. Frankel [67] assumes Ω to be constant over time. In light of the discussions of conditional heteroscedasticity in Section 3, this assumption now seems particularly strong although one possible argument in Frankel's defense is that misspecification of a model will frequently manifest itself in the form of conditional heteroscedasticity. Hence, if his model is correct, conditional heteroscedasticity may not be present. Frankel [67, p. 259] offers the following interpretation of (5.48):

The asset demands consist of two parts. The first term α represents the 'minimum-variance' portfolio. If an investor is extremely risk-averse ($\rho = \infty$), or if exchange rates are very uncertain ($|\Omega| = \infty$), the investor will hold countries' assets in the same proportions as the 'liabilities' represented by his consumption patterns. The second term represents the 'speculative portfolio'. A higher expected return on a given asset induces investors to hold more of that asset than is in the minimum-variance portfolio, to an extent limited only by the degree of risk-aversion and the uncertainty of the exchange rate. Again, in the case of risk-neutrality ($\rho = 0$), the assets become perfect substitutes and arbitrage insures that the expected relative returns are zero.

Inverting (5.48) gives a representation of the risk premiums that Frankel [67] investigates empirically,

$$E_t(\Delta s_{t+1}) + (i_t - i_t^\$ 1) = \rho\Omega(x_t - \alpha). \tag{5.49}$$

Since the nominal interest differential, $i_t^j - i_t^\$$, is equal to minus the logarithmic forward premium, $g_t^j - s_t^j$, for continuously compounded interest rates due to interest rate parity, the dependent variable in (5.42) is approximately $E_t(s_{t+1}^j - g_t^j)$ where the approximation is due to simple versus compound interest. The portfolio balance model, therefore, expresses the risk premium as a function of the optimal portfolio choices of individual investors.

Frankel [67] notes that estimation of (5.49) can proceed under the assumption of rational expectations and assumptions that allow aggregation such as identical tastes and homogenous information. With aggregation the portfolio shares become the actual shares of the respective assets in real wealth. He also notes that the nature of the approximations in (5.43) are of the same order of magnitude as the risk premium. His Appendix 1 repeats the deriation for the more rigorous definition of the real return, but the ultimate findings of the study are not affected by this change, so it is not reported here.

The econometric estimation of (5.49) substitutes the realized value of Δs_{t+1} minus its rational expectations prediction error for the conditional expectation to produce the following system of equations:

$$\Delta s_{t+1} + (i_t - i_t^{\$}1) = \rho\Omega(x_t - \alpha) + \varepsilon_{t+1}. \qquad (5.50)$$

Frankel [67] considers one of the major contributions of the paper to be estimation of the system of equations (5.50), subject to the restrictions imposed by his theory. To understand these restrictions recognize that the covariance matrix of the error process, ε_{t+1}, is the covariance matrix of unanticipated rates of depreciation, Ω. Hence, the coefficient matrix associated with x_t, $\rho\Omega$, is a scalar times the covariance matrix of the error process; and the vector of constants, $-\rho\Omega\alpha$, is minus the same scalar times the covariance matrix of the error process multiplied by the consumption shares which Frankel [67] measured from actual 1977 shares.

Appendixes 2 and 3 of Frankel [67] explain the maximum likelihood estimation of ρ and Ω subject to the constraints discussed above. The highly nonlinear nature of the restrictions complicates the analysis and requires solution of a system of quadratic equations.

The main result of Frankel's [67] estimation, which employs monthly data from July 1973 to August 1980, is the finding that the likelihood function is very flat, but it is decreasing in the relevant range $(\rho \geqslant 0)$; the maximum occurs at $\rho = 0$ implying that the representative international investor is risk neutral and that domestic and foreign assets are perfect substitutes.

Frankel [67] recognizes the limited nature of his findings. The failure to reject the null hypothesis does not imply that the null is true; the test may simply be not very powerful. Frankel also notes that tests of his model are tests of a composite hypothesis reflecting a number of simplifying assumptions. Typically, a composite hypothesis poses more of a problem when the null is rejected, since one is unsure which of the auxiliary assumptions, if any, is contributing to the rejection by the data; but a composite hypothesis can also lead to tests that lack power. Frankel notes that some of the simplifying assumptions include the nonstochastic nature of goods prices, the omission of alternative assets from the portfolio problem, and the assumed form of the utility function.

One restriction of the model that Frankel [67] does relax is the assumption of the same consumption shares across countries. His alternative version of the model can be aggregated in a similar way, and with the assumption of rational. expectations it produces the following system:

$$\Delta s_{t+1} + (i_t - i_t^\$ 1) = \rho \Omega (x_t - A w_t) + \varepsilon_{t+1}, \qquad (5.51)$$

where A is a matrix of consumption preferences whose columns represent the consumption shares across countries, and w_t is the vector of relative world wealths.

The results of this specification are not supportive of the model either, although in this case the maximized value of the likelihood function occurs somewhere above $\rho = 30$. Once again, though, the value of the likelihood function is very flat, and technically, a very wide range of parameter values including ρ equal to zero cannot be rejected by the data.

Incorporating stochastic inflation rates

Frankel and Engel [69] relax the assumption that the rates of inflation are predetermined variables. They define the real rate of return on the nominal asset of currency j by

$$(1 + r_{t+1}^j) = (S_{t+1}^j / S_t^j)(1 + i_t^j)(P_t^\$ / P_{t+1}^\$) \qquad (5.52)$$

where S_t^j is dollars per foreign currency, and the dollar price index is defined by

$$P_t^\$ \equiv \prod_{j=1}^{6} (P_t^j S_t^j)^{\alpha_j}. \qquad (5.53)$$

In (5.53) P_t^j is the consumer price index in country j, and α_j is the share of country j's goods in world consumption as measured by the ratio of its GNP relative to world GNP.

Frankel and Engel [69] write (5.42) as

$$W_{t+1} = W_t(x_t' z_{t+1} + 1 + r_{t+1}^\$) \qquad (5.54)$$

where $z_{t+1} = (r_{t+1} - r_{t+1}^\$ 1)$, the vector of differential real returns. The conditional expectation and conditional variance of W_{t+1} based on time t information are the following:

$$E_t(W_{t+1}) = W_t[x_t'E_t(z_{t+1}) + 1 + E_t(r_{t+1}^\$)] \qquad (5.55)$$

$$V_t(W_{t+1}) = W_t^2[x_t'V_t(z_{t+1})x_t + V_t(r_{t+1}^\$) + 2x_t'C_t(z_{t+1}; r_{t+1}^\$)]. \qquad (5.56)$$

If (5.55) and (5.56) are substituted into (5.47), and that equation is differentiated with respect to x_t, the result is the following system of equations:

$$E_t(z_{t+1}) = \rho C_t(z_{t+1}; r_{t+1}^\$) + \rho V_t(z_{t+1})x_t. \qquad (5.57)$$

Frankel and Engel [69] give empirical content to (5.57) by the assumptions that the conditional variance of z_{t+1} is a constant matrix Ω and that the vector of conditional variances is a constant. Under the additional standard assumption of rational expectations, the estimating system can be written as

$$z_{t+1} = \alpha + \rho \Omega x_t + \varepsilon_{t+1}, \qquad (5.58)$$

where $\Omega = E(\varepsilon_{t+1}\varepsilon_{t+1}')$.

Frankel and Engel [69] estimate (5.58) for five real currency returns relative to the US dollar. The five currencies are the Canadian dollar, the French franc, the Deutsche mark, the Japanese yen, and the UK pound. The sample period is June 1973 to August 1980, 87 monthly observations. Table XX presents the unconstrained OLS equation by equation estimation of (5.58) corresponding to their Table 1. Although each equation contains only one or two coefficients that are statistically significantly different from zero at conventional levels of significance, they report that the value of the log-likelihood function for the five equation system is 1086.49 while the value of the log-likelihood with all coefficients on the x_t variables equal to zero is 1057.11. Since twice the difference in the values of the log-likelihood functions is distributed as a chi-square variable with 25 degrees of freedom in this case, the hypothesis that all the coefficients are zero is rejected quite strongly. This occurs because the test statistic has a value of 58.76, while the 0.005 critical level for a chi-square with 25 degrees of freedom is 46.9278.

Frankel and Engel [69] also estimate the constrained model with and without estimation of a value for the coefficient of relative risk aversion, ρ. When ρ is constrained to be equal to 2, the maximized

TABLE XX
Unconstrained model from Frankel and Engel [69]

		Coefficients on									
Currency	Constants	$x_t^{c\$}$	x_t^F	x_t^{DM}	x_t^Y	$x_t^\$$	Log. lik.	R^2	D.W. Q(12)	S.E.R.	$F_{(5,81)}$
Canadian dollar	0.125^b (0.060)	-1.466^b (0.692)	-0.020 (0.322)	0.384 (0.243)	-0.120 (0.082)	0.150 (0.087)	251.87	0.09	2.04 25.32^b	0.01338	1.63
French franc	0.014 (0.138)	1.770 (1.584)	-1.132 (0.737)	-0.710 (0.557)	0.311 (0.188)	-0.159 (0.199)	179.87	0.08	2.37 6.86	0.03061	1.37
Deutsche mark	0.153 (0.145)	1.324 (1.669)	-0.818 (0.776)	-1.773^b (0.587)	0.361 (0.198)	-0.211 (0.210)	175.30	0.13	2.20 14.58	0.03226	2.48^b
Japanese yen	0.289^b (0.130)	0.319 (1.494)	-1.309 (0.695)	-2.213^b (0.525)	0.271 (0.177)	-0.141 (0.188)	184.98	0.21	2.04 12.26	0.02887	2.04
UK pound	0.028 (0.121)	1.772 (1.389)	-0.938 (0.646)	-0.993^b (0.488)	0.419^b (0.165)	-0.182 (0.175)	191.27	0.15	2.02 13.12	0.02685	2.86^b

Note: Dependent variable: $r_{t+1} - r_{t+1}^\$$, real rate of return on national currency relative to the dollar. Independent variables: x_t, shares of asset supplies in the world portfolio, with total wealth computed as the sum of the asset supplied, each evaluated at its respective exchange rate. Sample: June 1973 to August 1980 (87 observations). A superscript b signifies statistical significance at the 0.05 level.

value of the log-likelihood function is 1057.05. Since this is lower than the value when all coefficients are zero, it is clearly rejected relative to the unconstrained model. Frankel and Engel [69] choose to interpret this as a rejection of the optimization hypothesis, but this is clearly too strong given the number of auxiliary hypotheses that are necessary to conduct the empirical investigation. When ρ is allowed to be a free parameter, the maximum likelihood estimate of ρ is -67.0, and the value of the log-likelihood function is 1057.96. Since their model of a risk-averse investor only makes sense with ρ constrained to be greater than zero, this finding is a clear rejection of the model.

Potential reasons for rejection

Frankel and Engel [69] discuss several possible reasons for the rejection of their model. They note that investors may be rational maximizers of a function that depends only on the mean and variance of end-of-period wealth, as measured by their collection of assets, but these investors may face constraints that are not specified in the analysis. They also note that investors may be too sophisticated to maximize a function that depends only on the mean and variance of end-of-period wealth. Intertemporal asset pricing models collapse into such an environment only under relatively restrictive conditions, such as logarithmic utility or period-by-period independence of expected returns. Merton [162] notes that when the investment opportunity set is constant (i.e., the expected returns, the risk-free rate, and the covariance matrix of returns are constant), the portfolio decisions of an intertemporally maximizing investor are the same as those of a one-period, risk-averse, mean-variance investor. Unfortunately, Frankel [67] and Frankel and Engel [69] explicitly state that their frameworks are designed for macroeconomic applications when expected returns vary with the exogenous quantities of outside assets. Consequently, the one-period framework is suspect.

In their concluding remarks Frankel and Engel [69] also recognize that the estimation presumes a number of auxiliary assumptions, any one of which could be a source of the rejection of their model. In addition to the things mentioned directly above, they note that asset supplies must be properly measured from observations on government debt and foreign exchange intervention. They also mention the assumptions of constant conditional variances and covariances. One feature that they do not mention, but that is discussed in Frankel [67], is the role of other outside assets such as equity and the outside debt of third countries, much of which, but not all of which, is dollar denominated.

Unfortunately, it is difficult to know which of the assumptions in the Frankel and Engel [69] study is the most troublesome, but clearly, their model is too restrictive to explain the data.

Allowing for an error term

Several authors including Danker *et al.* [38], Lewis [135] and Rogoff [178] argue that portfolio balance specifications such as (5.48) are unlikely to hold without error. Allowing for an error term in (5.48) complicates the estimation considerably.

Rogoff [178] analyzes a Canadian dollar-US dollar uncovered interest rate parity equation in which deviations from uncovered interest rate parity depend on the ratio of Canadian dollar outside government bonds to the Canadian dollar value of US outside government bonds. Since the portfolio balance equation is postulated to hold with a serially correlated error, the inverted portfolio balance estimating equation analogous to (5.58) contains both the serially correlated portfolio balance error as well as the rational expectations error. Rogoff [178] estimates with two-step, two-stage least squares which is described above. He notes that an additional advantage of such an instrumental variable technique is its consistency under the alternative hypothesis that risk premiums are driven by relative government bond supplies and that sterilized intervention is effective. Frankel's strategy of treating bond supplies as exogeneous is inconsistent in this case.

Rogoff [178] employs weekly data and one-week forecasts from the period March 1973 to December 1980. He concludes [178, p. 147], "The data have resisted all our efforts to obtain equations for the uncovered interest rate differential in which portfolio balance variables appear with statistically significant coefficients of the right sign." He documents that the logarithmic forward rate

prediction error is serially correlated, thereby rejecting unbiasedness, but the relative bond supplies are powerless to explain the
serial correlation.

Such results are consistent with the findings in Danker et al. [38]
for Germany and Japan as well as Canada. Danker et al. carefully
construct series of outside nominal assets for these countries. They,
too, allow for a serially correlated error in the portfolio balance
equations. Their results for Germany indicate [38, p. 10], "clear
rejection of perfect substitutability (conditional on the assumption
of either rational or static expectations), but unfortunately they
offer little support for the alternative, the portfolio balance model."
The results for Japan and Canada are similarly inconclusive. One
potential inconsistency in the paper concerns the use of three month
interest rates with monthly data and the three-month rate of change
of the exchange rate. The estimation strategy outlined in their
Appendix A is appropriate for a one-month forecast. The nature of
their approximation with the forward filtering may deteriorate with
longer moving average processes.

Lewis [135] modifies Frankel's specification in two ways. She adds
an error term to the portfolio balance specification, and she
incorporates the covariance between exchange rates and inflation
rates. The estimation of the inverted bond demand functions
subject to the cross equation constraints requires simultaneous
estimation of a vector autoregression of the rates of change of the
exchange rates and the inflation rates. Lewis finds very small
estimates of the coefficient of relative risk aversion and strong
evidence against her model.

While the presence of a portfolio balance channel in the
determination of the risk premium cannot be ruled out on the basis
of these results, Lewis [135, p. 25] concludes, "The elusiveness of
the empirical relationship precludes the use of sterilized intervention as the basis for a finely-tuned or long-term exchange rate
policy."

Boothe et al. [19] similarly conclude that the evidence for a
portfolio balance channel is negligible. In contrast to others, they
conclude that the rejection of no-risk premium models is a rejection
of rational expectations. While such a conclusion is certainly a
possibility, the next section demonstrates that estimation of alternative models of the risk premium is feasible and potentially
enlightening.

5.2b. The representative consumer model

In Section 2 an intertemporal capital asset pricing model is presented. Mark [148] works directly with the first order conditions of the representative agent to examine the ability of this specification to capture time varying risk premiums in the forward foreign exchange market.

Consider equation (2.13) which is rewritten here for convenience in a slightly different form with a one good utility function:

$$E_t\{[\pi_{t+1}^m U'(C_{t+1})/\pi_t^m U'(C_t)][(S_{t+1} - G_t)/G_t]\} = 0, \qquad (5.59)$$

where π_{t+1}^m is the purchasing power of the dollar, $U'(C_t)$ is the marginal utility of consumption at time t, and G_t is the one month forward exchange rate. Mark [148] derives (5.59) by taking the difference between uncovered and covered investments in the nominally risk-free asset and dividing by the known foreign nominal return and by the constant discount factor that is known to the agent but not to the econometrician.

In order to proceed with estimation, Mark [148] specifies the representative agent's utility function to be the constant relative risk aversion form

$$U(C_t) = \delta C_t^{1-\gamma}/(1 - \gamma), \qquad (5.60)$$

where γ is the coefficient of relative risk aversion, $\gamma < 1$, and δ is an arbitrary constant.[43] By defining the ratio of consumption in period t to consumption in period $t + 1$ to be c_{t+1}, the ratio of the marginal utilities of consumption in (5.60) is c_{t+1}^γ.

Mark [148] uses monthly data from March 1973 to July 1983. He measures consumption with seasonally adjusted US consumption data from the *National Income and Product Accounts* of the Department of Commerce. He deflates the aggregate data by monthly estimates of population, obtained from the *Current Population Reports* of the Bureau of the Census, to derive a per capita series.

Data on US dollar foreign exchange rates were obtained from the Harris Bank's *Foreign Exchange Weekly Report*. The observations on S_{t+1} are sampled on the Friday closest to the end of the month,

---

[43] Empirical work with domestic US assets such as common stock and Treasury bills based on the representative agent model and the constant relative risk aversion utility function has been done by Grossman and Shiller [91], Hansen and Singleton [100, 101, 102], Dunn and Singleton [49], and Mehra and Prescott [161].

and the observations on G_t and S_t are taken four Fridays before that. The US dollar values of four currencies, the Canadian dollar, the Deutsche mark, the Dutch guilder, and the UK pound are used.

Mark [148] follows Hansen and Singleton [100, 101, 102] in employing two measures of consumption. Expenditures on non-durables (denoted ND) and on nondurables plus services (denoted NDS) are used.[44] The purchasing power of the dollar is measured by taking the inverse of the implicit price deflators for these consumption series.

The estimation strategy that Mark [148] employs is also Hansen's [96] GMM procedure which is discussed in Section 3. The orthogonality conditions are generated from (5.59). Under the null hypothesis, the composite variable inside the conditional expectation is orthogonal to the time t information set. Hence, as in Hansen and Singleton [100], estimation proceeds by specification of a vector of instruments from the time t information set. Mark [148] employs six sets of instruments. Each set includes the lagged consumption ratio, c_t, and either k lags, $k = 0$, 1, 2, of the four normalized profits, $(S_{t-k} - G_{t-k-1})/S_{t-k-1}$, or the four forward premiums, $(G_{t-k} - S_{t-k})/S_{t-k}$. The estimation therefore uses either 20, 36, or 52 orthogonality conditions to estimate the single parameter γ. As explained in Section 5.1, the product of T and the minimized value of the criterion function provides a chi-square statistic that tests the overidentifying restrictions of the model.

The results of Mark's [148] analysis are reported in Table XXI which reproduces his Tables 2 and 3. The point estimates of γ, denoted γ_T, are quite large and the standard errors encompass a wide range of possible values. This finding is typical of studies such as Hansen and Singleton [100, 101, 102] that employ a similar specification. The interesting aspect of the analysis is that there is not strong evidence against the model from the overidentifying restrictions. Only for the specification with no lags and the forward premiums as instruments do we find evidence against the model.

---

[44] See Eichenbaum and Hansen [53] for a discussion of alternative specifications of the representative agent model that incorporate service flows from durable goods and leisure in the utility function. Dunn and Singleton [49] also explore specifications that allow for durable goods. Miron [163] is unable to reject the representative agent specification when he allows seasonal influence on preferences and estimates with seasonally unadjusted data.

TABLE XXI
Representative agent model from Mark [148]

Consumption Specification	Instruments and Lags	γ_T	Std. Err. γ_T	χ^2 (df)	df	Confidence
NDS	I, 0	1.42	42.43	19.247	19	0.559
NDS	I, 1	39.98	29.08	41.270	35	0.785
NDS	I, 2	50.38	23.82	59.777	51	0.813
ND	I, 0	0.00	24.93	19.006	19	0.544
ND	I, 1	2.82	19.40	41.196	35	0.782
ND	I, 2	7.25	15.65	60.202	51	0.832
NDS	II, 0	43.51	33.13	32.301	19	0.972
NDS	II, 1	44.95	26.84	43.194	35	0.838
NDS	II, 2	41.45	23.65	56.450	51	0.721
ND	II, 0	17.51	26.25	33.227	19	0.977
ND	II, 1	13.79	18.97	44.377	35	0.867
ND	II, 2	12.67	16.64	57.079	51	0.741

Notes: The consumption specification uses either nondurables (ND) or nondurables plus services (NDS). Instruments always include k lags of the consumption ratio and either k lags of speculative profits (I) or the forward premiums (II), for $k = 0$, 1, 2. The chi-square statistic is T times the minimized value of the criterion function. Confidence is one minus the marginal level of significance of the test.

Given the strength of the evidence against unbiasedness that is reported in Fama [59] and Hodrick and Srivastava [106] employing the forward premiums as instruments, it appears that the representative agent model captures some aspects of the deviation of forward rates from expected future spot rates. As Mark [148, p. 16] states, "the evidence against the model is not overwhelming."[45]

Mark [148] outlines several caveats to the modeling that involve serious but difficult extensions of the analysis. Hansen and Singleton [100] note that identification of such a model occurs because the enconometrician assumes that none of the sources of uncertainty is from the utility function of the representative agent. Garber and King [81] argue that this is particularly implausible and demonstrate how such 'Euler equation estimation' may lead to 'mongrel' parameters.

---

[45] Obstfeld [169] explores the implications of the representative agent model with consumption data from the US, Germany and Japan. He notes that if agents in each country can trade the nominally risk-free asset, then the differences in the nominal intertemporal marginal rates of substitution defined in the same currency must be zero. He also finds evidence supporting the model.

A second caveat concerns the simple specification of the utility function. Mehra and Prescott [161] argue that the simple representative agent model is inconsistent with the observed average risk premium in the returns on stocks over returns on treasury bills. Eichenbaum and Hansen [53] explore whether alternative time-nonseparable specifications are capable of generating the type of fluctuations in the intertemporal marginal rates of substitution that are necessary to generate consistency of the representative agent model with data. The representative agent model may or may not be capable of such reconciliations.

Lucas [146] argues that the econometric examination of more and more complicated specifications of the representative agent model will almost surely result in statistical rejection of the model along some dimension. It can be argued that economic models, especially ones that are analytically tractable, are obviously false descriptions of a complicated world. They guide our thinking about the way the world works and allow us to understand the outcomes of relatively complicated experiments. Requiring too much consistency with the data in a classical statistical framework may actually hinder our understanding of the way the world works. In this respect the work of Korajczyk [122] and Mark [148] provides us with an interpretation of the rejection of the unbiasedness hypothesis that is not overwhelmingly at variance with the data.

These explanations are consistent with a rational world. If the rejection of the unbiasedness hypothesis were due to inadequate speculation or an inefficient market, it seems unlikely that the data would have provided as much support for this model as is found.

6. EVALUATION OF FORECASTS

Previous sections have examined the forward rate as a predictor of the future spot rate and various models of risk premiums that might be used to adjust the forward rate to improve its forecasting ability. In this section some consideration is given to analysis of advisory services that sell forecasts of exchange rates and some survey evidence on expectations of future exchange rates. The section concludes with a review of a forecasting experiment based on Kalman filtering.

6.1. Advisory service forecasts

In a sequence of interesting articles for *Euromoney*, Levich [131, 132] analyzed the accuracy of several advisory services that sell forecasts of exchange rates to corporate customers. Levich [131] notes that if forward markets are characterized by time-varying risk premiums, then advisory services ought to be able to outperform the forward rate. It also follows that any profits earned from trading on the forecasts may simply reflect fair compensation for bearing exchange risk rather than an excess return for exploiting unusual forecasting expertise.

Levich [131] also notes that the evaluation of forecasts is a tricky issue. The size of forecast errors may be irrelevant for some decision makers who prefer essentially a directional prediction. For these individuals, mean error or mean squared error is irrelevant if the predicted direction of forecast is correct.

Statistical valuation of the forecasts is also complicated by the fact that most experiments are done on US dollar denominated exchange rates. Hence, if a service is forecasting the US dollar value of nine currencies, the realizations are not nine independent draws from a normal probability table. Also, for several of the services, only a year or two of data were available.

Several findings of the Levich [131, 132] studies provide confirmation of what one intuitively expects in a volatile asset market. The forecasting ability (as measured by the percentage of correct directional forecasts) of the advisory services changes from year to year and is not uniform across currencies or maturities. Most services that do well predict no more than around 55 to 60 percent correctly and rarely more than 70 percent in a given year. In fact, for the period 1978–1982 at the one month horizon, there are only five statistically significant forecasts at the 0.15 marginal level of significance out of 90 observations, nine currencies and ten advisory services. As a group, the advisors are far from statistically significant.

When the results are pooled across all currencies, the results are even less supportive with only three agencies having greater than fifty percent correct forecasts with only one agency providing statistically significant performance at the 0.14 marginal level of significance.

The results are not appreciably better at longer horizons where the lack of nonoverlapping data hampers inference. Since the percentage of correct forecasts must be higher to be judged statistically significant with a smaller number of observations, and because the observations are not strictly independent, it is difficult to conclude that any forecasting service has significant forecasting ability.

The deterioration in the quality of forecasts during 1982 caused Levich [132, p. 146] to close his analysis by stating, "The nagging possibility is that foreign exchange relationships may evolve more quickly than many forecasters can comprehend and make adjustments." The strength of the dollar during the early 1980's was indeed surprising to most economists and unpredicted by most economic models.

Goodman and Jaycobs [87] analyze the forecasting performance of advisory services that sell forecasts based on technical analysis. These technical models are more sophisticated counterparts to the filter rules discussed earlier. They typically base their recommendations solely on phenomena in exchange markets such as relative rates of change, moving averages of past prices or momentum, but also on pattern recognition of waves and cycles, as well as other market factors such as open interest.

The profitability of the strategies is described by expressing the realized profit as a percentage of the forward price which is a return on capital at risk. For the sixteen services that forecast the dollar values of six major currencies, for the period from January 1980 to December 1982, the average returns ranged from 6.6% to 19.0% on a per annum basis. One interesting feature of the results is a correlation between the return and the turnover in the predictions. The highest returns were earned by the technical models that change their predictions approximately once a month. If time-varying risk premiums characterize the foreign exchange market, technical models such as are described above may be able to determine movements in the risk premium. An alternative explanation is, of course, that the technical models are capturing a market inefficiency.

6.2. Results using survey expectations

Frankel and Froot [70] and Froot and Frankel [79] employ survey data of expectations of future exchange rates to investigate various aspects related to the efficiency of the forward market. The surveys are taken from several sources.

The Amex Bank Review (AMEX) publishes surveys taken once or twice annually from early 1976 to the present although apparently no surveys were conducted during 1979 or 1980. For each survey, 250–300 central bankers, private bankers, corporate treasurers, finance directors and economists are asked to record their expectations of five major currencies against the dollar for 6 months into the future. The five currencies are the French franc, the Deutsche mark, the Swiss franc, the Japanese yen, and the UK pound.

A second source of survey data is the Economist Financial Report (ECON) that began conducting surveys in June 1981 of 13 leading international banks. Six times annually respondents are polled for their expectations at three and 6-month horizons for the same five currencies as AMEX.

These first two sources of survey data are employed by Frankel and Froot [70]. In addition, Froot and Frankel [79] employ survey data from Money Market Services, Inc. (MMS). Every two weeks, between January 1983 and October 1984, MMS polled by telephone an average of 30 currency traders or currency-room economists at major international banks for their expectations of the dollar value of the UK pound, the Deutsche mark, the Swiss franc, and the Japanese yen at a two-week horizon and a one-month horizon.

Two obvious problems plague the use of these data. First, many economists are justifiably suspicious of surveys preferring instead an approach based on positive economics. Frankel and Froot [70, p. 15] note, "a cornerstone of 'positive economics' is that we learn more by observing what people do (in the marketplace) than what they say." Consider the incentive problem of a trader who possesses private information that he has used to construct a portfolio of positions based on the deviations of his expectations from the current forward rates. When MMS calls him for his expectations, will he reveal his information, or will he lie and quote something like the forward rate? Does he even know that the two are different?

A second problem concerns the length of the time series. The AMEX data contain 11 nonoverlapping observations. The ECON data have 16 nonoverlapping 3-month observations and eight nonoverlapping 6-month observations. The MMS data have 34 nonoverlapping 2-week observations and 17 1-month observations.

With such small numbers of degress of freedom and short time intervals, and given the volatility of government policies during the 1970's and 1980's, the application of bootstraps and large sample distribution theory may be tenuous.

Indeed, Frankel and Froot [70] find strong evidence of unconditional bias in both the forward premium as a predictor of depreciation and in the survey expectations of depreciation. While admitting that either a risk premium or a gross departure from rational expectations could explain the former finding, an attractive alternative is that it is consistent with both, that rational expectations of the value of the US dollar relative to foreign currencies during this period had to assess several possible scenarios for the relevant monetary and fiscal policies as well as comparative growth prospects for the countries. If these policies have not all been realized during the sample, the statistics would be untrustworthy. Although Frankel and Froot (1985) are aware of the small number of degrees of freedom and conserve them by estimating single parameters in equations that group several currencies, their use of bootstrap techniques suffers from the same criticism that was discussed in regard to Korajczyk's [122] study.

Froot and Frankel [79] use a GMM estimator that assumes conditional homoscedasticity to examine the survey data. It is interesting to note that evidence of unconditional bias in either the forward premium or the survey expections tends to be concentrated only in the data from short time intervals or with small numbers of nonoverlapping intervals. Regressions of the forecast error from the survey data on a constant and the forward premium generally indicate rejection of the hypothesis that the survey data of expected depreciations are the rational conditional expectations of the actual rates of depreciation. Such a finding is similar to Brown and Maital's [27] findings that the Livingston survey data on inflation, 16 years of 6-month forecasts, are generally unbiased but do not make full use of information available when the forecast is made. While such results are potentially troublesome for the rational expectations hypothesis, they may simply reflect the problems inherent in the use of survey data mentioned above. They may also be indicative of the problems encountered in doing inference with economic time series that are relatively short in comparison to the length of time that some economic events require to be observed.

If the findings of Frankel and Froot [70] and Froot and Frankel [79] are supported with additional data, they will have overturned one of the findings discussed above. With direct observations on expectations of depreciation, they are able to examine the variability of expected rates of depreciation and of risk premiums. Their findings indicate that, in contrast to the discussions of Fama [59] and Hodrick and Srivastava [108], the variability of expected rates of depreciation is larger than the variability of risk premiums. Only additional data will allow us to ascertain which methodology, if either, is correct, and which inferences, if any, is to be believed.

6.3. A Kalman filter forecasting approach

The unobservable nature of risk premiums and expected rates of depreciation led Wolff [217] to postulate a state-space model given by the following two equations:

$$g_t - s_{t+1} = p_t + \varepsilon_{t+1} \qquad (6.1)$$

$$p_t = \alpha p_{t-1} + \eta_t \qquad (6.2)$$

where p_t is the logarithmic risk premium, ε_{t+1} is the innovation in the logarithm of the exchange rate, s_{t+1}, and η_t is the innovation in the risk premium. The Kalman filter can be applied to this system by viewing (6.1) as the observer equation and (6.2) as the state transition equation. Application of the Kalman filter requires specification of two parameters, α and the ratio of the variance of η_t to the variance of ε_{t+1}.

Wolff [217] searches over a range of possible values for these parameters varying α from 0.00 to 1.10 with 0.01 increments and varying the ratio of the variances from 10^{10} to 0 with decrements of 0.1. The exchange rates in his study are the US dollar values of the UK pound, the Deutsche mark, and the Japanese yen. The data are from the Harris Bank Data Base, Friday closing observations on 30-day forecasts, from 6 April 1973 to 13 July 1984.

The metric Wolff [217] employs in evaluating the ability of the Kalman filtering experiment to improve the forecast of the forward rate is a modification of Thiel's [208] U-statistic, the ratio of the root mean-square error of the forecast to the root mean-square error of the naive forecast of no change. Essentially, the naive measure asserts that the logarithm of the spot rate follows a random walk.

Wolff [217] finds that the state space model produces better forecasts than the forward rate over a wide range of parameter values, but in no case does the model produce a smaller U-statistic than one, the value under the random walk assumption. These best forecasting models have highly autocorrelated risk premiums and innovation variances in exchange rates that are between 10 and more than 100 times as variable as the innovations in risk premiums implying that between nine and 16 percent of the variability of the forward rate forecast errors is due to risk premiums.

One disturbing feature of these results is that the adjustments of the Kalman filtering model never produce forecasts that dominate the spot rate. While there are no standard errors associated with the point estimates, if the approach is truly capturing a risk premium, it ought to be capable of beating the no change prediction. Perhaps the structure of the state-space model is simply too rigid and does not employ sufficient other economic information to produce such forecasts.

7. EMPIRICAL INVESTIGATION OF FOREIGN CURRENCY FUTURES

This section examines some empirical investigations of foreign currency futures prices. It begins with a discussion of the empirical analysis of the difference between futures prices and forward prices that is conducted in Cornell and Reinganum [33]. Then, the evidence of Hodrick and Srivastava [107] and McCurdy and Morgan [154] on unbiasedness of daily futures prices is examined. The section concludes with a discussion of Feinstone's investigation of intraday price movements.

7.1. Forwards versus futures

Section (2.2) explains the institutional features of the foreign currency futures market and presents the Cox, Ingersoll and Ross [34] arbitrage argument that equates the futures price to the present value of a stochastic quantity of foreign exchange equal to the product of the short-term interest rates between the date of the contract and the date of delivery. The difference between the forward price and the futures price at time t for delivery at time $t + k$ is given in (2.27).

Cornell and Reinganum [33] examine the difference between these two prices with data suppplied by the International Monetary Market of the Chicago Mercantile Exchange, which is commonly referred to as the IMM. Foreign currency futures contracts on the IMM specify delivery on the third Wednesday of March, June, September and December. Trading stops on the Monday preceding the third Wednesday if Monday, Tuesday and Wednesday are legitimate business days. This is consistent with the 2-day delivery procedure in the spot market.

Their comparison uses data from 21 contracts beginning with the June 1974 contract and ending with the June 1979 futures contract. The currencies studied are the UK pound, the Canadian dollar, the Deutsche mark, the Japanese yen, and the Swiss franc. Four different maturities are examined. They are 1, 2, 3 and 6 months prior to delivery.

One problem preventing a completely precise comparison arises because the futures price data are closing prices on the IMM (closing occurs between 1:15 p.m., Chicago time, for the Swiss franc and 1:25 p.m. for the Japanese yen), while the forward rates, which were provided by the Continental Illinois Bank, are measured at 1:00 p.m., Chicago time. Cornell and Reinganum note that this produces some random variation in the two prices that is not present in the theory.

Cornell and Reinganum [33] examine whether the mean deviations between the forward prices and the futures prices are statistically significantly different from zero. Of the 20 mean values, only the value for the 3 month UK pound obtains a marginal level of significance smaller than 0.01. Only one other contract, the 1-month Canadian dollar, has a mean value large enough to be statistically significantly different from zero at the 0.05 level.

Cornell and Reinganum [33] note that existence of a zero mean does not preclude arbitrage profits. They examine the data for the three month Deutsche mark and the three month Swiss franc, observation by observation, finding that the maximum discrepancy is 17 points or about three times the bid-ask spread in the forward market. Since the data are not measured precisely at the same time, this is not a large difference.

Cornell and Reinganum [33] conclude their study with an examination of the covariances between the percentage changes in the daily foreign currency futures prices and the percentage changes in bid prices for Treasury bills maturing on the delivery day. The covariances are estimated from a sample of 40 trading days beginning 60 days prior to the maturity date. All of the covariances are very small, on the order of 1.0×10^{-7}. With such a small covariance, they demonstrate that the empirical results on the differences between the futures prices and forward prices are consistent with the implications of the model in Cox, Ingersoll and Ross [34].

Section (2.2) also demonstrates that futures prices may be biased predictors of future spot prices if the future spot rate covaries with the intertemporal marginal rate of substitution times the product of the short interest rates.

7.2. Testing unbiasedness of daily futures

Hodrick and Srivastava [107] argue that if there is little difference between forward prices and futures prices of foreign currencies, and if the unbiasedness hypothesis can be rejected for forward rates, then consistency of the data requires that the futures prices must be biased as well. They also demonstrate that in such a world a futures price will not be an unbiased predictor of the futures price on the following day.

Samuelson [188] presents an argument to indicate how the sequence of futures prices in an efficient market can be a martingale with $F_{t,k} = E_t(F_{t+1,k-1})$. To see that this implies unbiasedness of futures prices as predictors of future spot rates, apply the daily unbiasedness hypothesis for k days and recognize that $F_{t+k,0} = S_{t+k}$ by arbitrage. Then apply the iterated expectations argument k times. Since the time t information set is a proper subset of the $t + 1$ information set,

$$F_{t,k} = E_t(F_{t+1,k-1}) = E_t[E_{t+1}(F_{t+2,k-2})] = E_t(F_{t+2,k-2}). \quad (7.1)$$

Clearly, if forward prices and futures prices are essentially identical, applying the unbiasedness test to futures prices with monthly data, would provide a rejection of that hypothesis. Reconciliation of Samuelson's [188] argument and the data only requires that one recognize that Samuelson's [188] argument is built on the hypothesis of risk neutrality.

Hodrick and Srivastava [107] demonstrate that if the daily intertemporal marginal rate of substitution of money is stochastic, futures prices will generally be equal to the conditional expectation of the fututes price on the next day plus a risk premium.

From (2.7) notice that $Q_{t+k,k}^m = Q_{t+k,k-1}^m Q_{t+1}^m$. With this result, it is easy to demonstrate that

$$F_{t,k} = E_t(F_{t+1,k-1}) + C_t(F_{t+1,k-1}; Q_{t+1}^m R_t). \qquad (7.2)$$

Whenever the conditional covariance in (7.2) is not constant, there will be time varying bias in daily futures prices.

Hodrick and Srivastava [107] investigate daily unbiasedness with empirical techniques analogous to those discussed in Section (3.7). The unbiasedness hypothesis implies

$$(F_{t,k} - S_t)/S_t = E_t[(F_{t+1,k-1} - S_t)/S_t], \qquad (7.3)$$

where the transformation is legitimate because S_t is in the time t information set. Such a transformation makes it more likely that the data will satisfy the assumption of stationarity necessary for convergence of the asymptotic distributions of the estimated parameters.

Hodrick and Srivastava [107] follow Jagannathan [116, 117] in deriving the asymptotic distribution of their GMM estimators. When using data from futures markets, asymptotic distributions must be derived by allowing the number of contracts to go to infinity. The stationary stochastic process can be considered to be a hypothetical vector of daily observations on the futures premiums on the left-hand side of (7.3) with between 1 and K days to maturity, where K is the longest maturity in the actual data. The observations on actual data are a particular sampling from this hypothetical vector, and the variances and covariances used in the study are averages of the covariances of the hypothetical vector.

Hodrick and Srivastava [107] also employ a system estimation technique that is a GMM analogue to Zellner's [219] seemingly unrelated regression that does not impose an assumption of conditional homoscedasticity.

TABLE XXII
Full sample system estimates from Hodrick and Srivastava [107]

$$(F^j_{t+1,k-1} - S^j_t)/S^j_t = \alpha_j + \beta_j[(F^j_{t,k} - S^j_t)/S^j_t] + \varepsilon^j_{t+1}, \quad j = 1, \ldots, 5$$

Currency	$\hat{\alpha}_j$	$\hat{\beta}_j$	R^2	Test $\beta_j = 1$	SEE
1. Japanese yen	0.0001 (0.0001) [0.3095)	0.9303 (0.0139) [0.0000]	0.72	25.3290 [0.0000]	0.0063
2. Deutsche mark	0.0005 (0.0001) [0.0007]	0.8665 (0.0195) [0.0000]	0.45	47.0878 [0.0000]	0.0063
3. UK pound	−0.0003 (0.0001) [0.0197]	0.9258 (0.0170) [0.0000]	0.58	19.1229 [0.0000]	0.0059
4. Canadian dollar	−0.0001 (0.0001) [0.0074]	0.9395 (0.0169) [0.0000]	0.62	12.8669 [0.0003]	0.0024
5. Swiss franc	0.0007 (0.0002) [0.0000]	0.8940 (0.0171) [0.0000]	0.53	38.5584 [0.0000]	0.0075

Notes: The sample period is 6/1/73 to 12/8/83 and contains 2420 observations. Standard errors are in parenthesis, and marginal levels of significance are in brackets. The test statistic is $[(1 - \hat{\beta})/\sigma(\hat{\beta})]^2$ where $\sigma(\hat{\beta})$ is the standard error of the estimate. The test statistic is distributed as a chi-square with one degree of freedom. Test all $\beta_j = 1$, $\chi^2(5) = 81.2868$ [0.000]. Test of equality of β_j, $\chi^2(4) = 11.8578$ [0.0184]. Test of overidentifying restrictions, $\chi^2(20) = 54.6413$ [0.0000].

The futures prices in the studies are the US dollar values of the Japanese yen, the Deutsche mark, the UK pound, the Canadian dollar and the Swiss franc. The sample is 1 June 1973 to 8 December 1983. There are 2420 daily observations in the full sample.

The system estimation is presented in Table XXII, which reproduces Table 7 of Hodrick and Srivastava [107]. In each case the null hypothesis that $\beta_j = 1$, which allows for a constant but not a time varying risk premium, is rejected by the data at all marginal levels of significance larger than 0.0003. The coefficients range from a low of 0.8665 for the Deutsche mark to 0.9395 for the Canadian dollar. Because the standard errors are quite small, the null hypothesis that the coefficients are equal has a marginal level of significance of 0.0184. The test of the overidentifying restrictions of the model, which tests the joint hypothesis that the other four futures premiums would not have been useful in predicting a particular $(F^j_{t+1,k-1} - S^j_t)/S^j_t$, is a chi-square statistic with 20 degrees of freedom. The value of the statistic is 54.6413 which is larger than the critical level associated with the 0.001 marginal level of significance.

Hodrick and Srivastava [107] also express concern that their findings may be dependent on the use of data from the early part of the experience with floating exchange rates. It is often argued in popular discussions of market efficiency that prices in speculative markets such as futures and options possess stochastic properties that correspond to theories of efficient markets only after the markets are mature.

Consequently, Hodrick and Srivastava [107] split the sample into two equal subsamples. The results of the system estimation for subsample one are presented in Table XXIII which reproduces Table 9 from Hodrick and Srivastava [107].

TABLE XXIII
Subsample one system estimates from Hodrick and Srivastava [107]

$$(F^j_{t+1,k-1} - S^j_t)/S^j_t = \alpha_j + \beta_j[(F^j_{t,k} - S^j_t)/S^j_t] + \varepsilon^j_{t+1}, \qquad j = 1, \ldots, 5$$

Currency	$\hat{\alpha}_j$	$\hat{\beta}_j$	R^2	Test $\beta_j = 1$	SEE
1. Japanese yen	0.0002 (0.0001) [0.0538]	0.9509 (0.0175) [0.0000]	0.87	7.8556 [0.0051]	0.0030
2. Deutsche mark	0.0006 (0.0002) [0.0006]	0.8279 (0.0330) [0.0000]	0.36	27.1591 [0.0000]	0.0038
3. UK pound	−0.0002 (0.0002) [0.1973]	0.9553 (0.0234) [0.0000]	0.66	3.6520 [0.0560]	0.0035
4. Canadian dollar	−0.0002 (0.0001) [0.0080]	0.9337 (0.0192) [0.0000]	0.73	11.9104 [0.0006]	0.0014
5. Swiss franc	0.0007 (0.0002) [0.0012]	0.8969 (0.0298) [0.0000]	0.47	11.9732 [0.0005]	0.0044

Notes: See Table XXII. Test of all $\beta_j = 1$, $\chi^2(5) = 48.3672$ [0.0000]. Test of equality of β_j, $\chi^2(4) = 12.8075$ [0.0123]. Test of overidentifying restrictions, $\chi^2(20) = 55.6371$ [0.0000].

TABLE XXIV
Subsample two system estimates from Hodrick and Srivastava [107]

$$(F_{t+1,k-1}^j - S_t^j)/S_t^j = \alpha_j + \beta_j[(F_{t,k}^j - S_t^j)/S_t^j] + \varepsilon_{t+1}^j, \qquad j = 1, \ldots, 5$$

Currency	$\hat{\alpha}_j$	$\hat{\beta}_j$	R^2	Test $\beta_j = 1$	SEE
1. Japanese yen	0.0007 (0.0003) [0.0152]	0.8562 (0.0262) [0.0000]	0.43	30.1638 [0.0000]	0.0056
2. Deutsche mark	0.0004 (0.0002) [0.4239]	0.8636 (0.0250) [0.0000]	0.46	29.7165 [0.0000]	0.0050
3. UK pound	−0.0002 (0.0002) [0.2510]	0.8664 (0.0274) [0.0000]	0.40	23.8227 [0.0000]	0.0047
4. Canadian dollar	−0.0001 (0.0001) [0.2867]	0.9492 (0.0298) [0.0000]	0.50	2.9132 [0.0879]	0.0020
5. Swiss franc	0.0010 (0.0003) [0.0018]	0.8731 (0.0223) [0.0000]	0.51	32.4438 [0.0000]	0.0061

Notes: See Table XXII. Test of all $\beta_j = 1$, $\chi^2(5) = 59.9767$ [0.0000]. Test of equality of β_j, $\chi^2(4) = 6.8491$ [0.1441]. Test of overidentifying restrictions, $\chi^2(20) = 30.2505$ [0.0659].

Once again, there is strong evidence against the null hypothesis $\beta = 1$ for all currencies except for the UK pound which has an individual marginal level of significance of 0.056. All other currencies have test statistics whose marginal levels of significance are smaller than 0.0051. The joint test that the five β's are equal to one has a marginal level of significance smaller than 0.0001.

The results for subsample two are qualitatively quite similar. They are presented in Table XXIV which reproduces Table 11 of Hodrick and Srivastava [107]. In the individual tests of $\beta = 1$ only the Canadian dollar has a test statistic whose marginal level of significance is as large as 0.0879. The other four currencies have values of the test statistic that are larger than the critical level associated with the 0.0001 significance level. Not surprisingly, the joint test that $\beta_j = 1$ for all five currencies also has a test statistic that is larger than the 0.0001 critical level.

In the first subsample there is strong evidence that the set of excluded futures premiums may have been useful in predicting the dependent variables since the value of the test statistic for the twenty overidentifying restrictions is 55.6371 which is larger than the critical level of a chi-square with twenty degrees of freedom associated with the 0.0001 level of significance. For the second subsample this strength of evidence that other data would have been useful in predicting the dependent variables is not present since the test of the overidentifying restrictions has a value of 30.2505 which corresponds to a marginal level of significance of 0.0659.

From Section (4), equation (6.2) can be written as

$$F_{t,k} = E_t(F_{t+1,k-1}) + P_t \qquad (7.4)$$

where $P_t \equiv C_t(F_{t+1,k-1}; Q^m_{t+1}R_t)$ is the one-day risk premium. This is the expected profit on a short position in the foreign currency futures market. The transformation of variables in (7.3) produces a normalized risk premium,

$$(F_{t,k} - S_t)/S_t = E_t[(F_{t+1,k-1} - S_t)/S_t] + p_t \qquad (7.5)$$

where $p_t \equiv P_t/S_t$. Then, the slope coefficients in the daily data regressions discussed above can be decomposed as

$$\beta = C[(F_{t+1,k-1} - S_t)/S_t; (F_{t,k} - S_t)/S_t]/V[(F_{t,k} - S_t)/S_t]$$
$$= \{V[E_t(F_{t+1,k-1} - S_t)/S_t]$$
$$+ C[E_t((F_{t+1,k-1} - S_t)/S_t); p_t]\}/V[(F_{t,k} - S_t)/S_t]. \qquad (7.6)$$

The average estimated value of β in Table XXIII is approximately 0.9, and this provides evidence that $V[E_t((F_{t+1,k-1} - S_t)/S_t)]$ is greater than $V(p_t)$. To establish this result subtract one from two times β which expresses the difference between the variances of $E_t[(F_{t+1,k-1} - S_t)/S_t]$ and p_t as a proportion of the variance of their sum. The calculation is

$$2\beta - 1 = \{V[E_t((F_{t+1,k-1} - S_t)/S_t)] - V(p_t)\}/V[(F_{t,k} - S_t)/S_t] = 0.8 \qquad (7.7)$$

with a standard error of approximately 0.04. The data indicated substantially greater variability in $E_t[(F_{t+1,k-1} - S_t)/S_t]$ than in p_t.

This result contrasts sharply with the evidence discussed above from Fama [59] and Hodrick and Srivastava [108] with monthly data and the one-month forward premium. The studies using monthly forward exchange rate data indicate substantially greater variability in the risk premium component of the forward premium than in the expected rate of depreciation.

Hodrick and Srivastava [107] discuss a possible way to reconcile the two findings. The analysis is facilitated by consideration of the logarithmic counterpart to (7.5) where lower case letters represent natural logarithms of their upper case counterparts:

$$f_{t,k} - s_t = E_t(f_{t+1,k-1} - s_t) + p_t, \tag{7.8}$$

and where p_t now represents the logarithmic risk premium as in Fama [59]. A regression of $f_{t+1,k-1} - s_t$ on a constant and the logarithmic futures premium defined on the left-hand side of (7.8) produces a slope coefficient β in (7.6) that is denoted β_1 to represent a one-day forecast.

Hodrick and Srivastava [107] note that expanding to a two-day forecast interval as in

$$f_{t+2,k-2} - s_t = \alpha_2 + \beta_2(f_{t,k} - s_t) + \varepsilon_{t+2,2} \tag{7.9}$$

produces

$$\beta_2 = \beta_1 - \{C[(p_t; E_t(p_{t+1})]$$
$$+ C[E_t(f_{t+1,k-1} - s_t); E_t(p_{t+1})]\}/V(f_{t,k} - s_t). \tag{7.10}$$

The result in (7.10) follows because the dependent variable can be decomposed into

$$f_{t+2,k-2} - s_t = f_{t+2,k-2} - f_{t+1,k-1} + f_{t+1,k-1} - s_t \tag{7.11}$$

and $E_t(f_{t+2,k-2} - f_{t+1,k-1}) = -E_t(p_{t+1})$. Generalizing (7.10) to H-day forecasts produces

$$\beta_H = \beta_1 - \left\{ C\left[p_t; \sum_{h=1}^{H} E_t(p_{t+h})\right] \right.$$
$$\left. + C\left[E_t(f_{t+1,k-1} - s_t); \sum_{h=1}^{H} E_t(p_{t+h})\right] \right\} \bigg/ V(f_{t,k} - s_t). \tag{7.12}$$

Hodrick and Srivastava [107] argue that if daily risk premiums are highly positively autocorrelated and not particularly negatively correlated with $E_t(f_{t+1,k-1} - s_t)$, then increasing the forecast inter-

TABLE XXV
Monthly data system estimates from Hodrick and Srivastava [107]

$(F^j_{t+30,k-30} - S^j_t)/S^j_t = \alpha_j + \beta_j[(F^j_{t,k} - S^j_t)/S^j_t] + \varepsilon^j_{t+30}, j = 1, \ldots, 5.$

Currency	$\hat{\alpha}_j$	$\hat{\beta}_j$	R^2	Test $\beta_j = 1$	SEE
1. Japanese yen	0.0001 (0.0026) [0.9752]	0.4552 (0.1377) [0.0009]	0.03	15.6610 [0.0000]	0.0317
2. Deutsche mark	−0.0020 (0.0029) [0.4944]	0.5613 (0.2382) [0.0184]	0.03	3.3927 [0.0655]	0.0338
3. UK pound	−0.0059 (0.0023) [0.0091]	0.3691 (0.1868) [0.0482]	0.03	11.4073 [0.0007]	0.0256
4. Canadian dollar	−0.0035 (0.0010) [0.0003]	0.0786 (0.2247) [0.0727]	0.00	16.8086 [0.0000]	0.0122
5. Swiss franc	0.0036 (0.0037) [0.3365]	0.1875 (0.1832) [0.3061]	0.00	19.6647 [0.0000]	0.0417

Notes: There are 115 nonoverlapping observations. Also, see Table XXII. Test of all $\beta_j = 1$, $\chi^2(5) = 46.5306$ [0.0000]. Test of equality of β_j, $\chi^2(4) = 4.9459$ [0.2929]. Test of overidentifying restrictions, $\chi^2(20) = 28.128$ [0.1064].

val causes β_H to fall as the variability of the risk premium increases relative to the variability of the expected rate of depreciation.

Hodrick and Srivastava [107] provide some limited empirical support for this interpretation by investigating a monthly forecast. Table XXV reproduces their Table 12 which is a system estimation of

$$(F_{t+30,k-30} - S_t)/S_t = \alpha + \beta(F_{t,k} - S_t)/S_t + \varepsilon_{t+30,30}. \quad (7.13)$$

There are 115 nonoverlapping observations in the analysis. A comparison of Table XXII and Table XXV indicates that all of the slope coefficients have fallen substantially. The slope coefficient for the Canadian dollar falls from 0.94 to 0.08 while the one for the Swiss franc falls from 0.89 to 0.19.

The variance decomposition in (7.7) now indicates greater variability of one-month risk premiums than of expected rates of depreciation as is also indicated by the monthly forward market data. While the consistency of these results is predicted by the approximate equality of forward rates and futures prices noted above, it does not actually demonstrate that the daily risk premiums have the requisite time series properties attributed to them in the discussion following (7.12). This appears to be a fruitful area for additional research.

McCurdy and Morgan [154] reexamine the conclusions of Hodrick and Srivastava [107] using the newly available data from the Center for Research in Futures Prices of the University of Chicago. Rather than employing a heteroscedasticity consistent estimation strategy, they model the conditional heteroscedasticity using the generalized autoregressive conditional heteroscedasticity (GARCH) process developed by Bollerslev [17].

The specification of their estimating equation also differs since they utilize a logarithmic transformation of the data in order to avoid dividing both sides of the equation (7.2) by a common stochastic variable. Their approach is to estimate the system

$$f_t - f_{t-1} = \gamma_0 + \gamma_1(f_{t-1} - f_{t-2}) + \gamma_2 M_1 + \gamma_3 M_2 + \gamma_4 W_1 + \varepsilon_t \quad (7.14)$$

where the null hypothesis is $f_{t-1} = E_{t-1}(f_t)$ implying that all of the γ coefficients are zero in (7.14). The error term ε_t is assumed to be conditionally heteroscedastic with conditional variance h_t that is modeled as

$$h_t = \alpha_0 + \alpha_1 \varepsilon_{t-1}^2 + \beta_1 h_{t-1}, \quad (7.15)$$

and nonnegativity constraints are imposed on the parameters of (7.15). The M_i variables in (7.14) are dummy variables that are unity prior to and after 1 October 1981 for Mondays and W_1 is a dummy that is unity prior to that date for Wednesdays. The use of dummy variables for days of the week is motivated by an acceleration in the clearing process for US dollar denominated cheques for purchases of foreign currency that occurred on 1 October 1981.

Table XXVI presents evidence from McCurdy and Morgan's [154] Table 1. Notice that under the null hypothesis the γ_1 coefficient should be zero in the logarithmic specification and should be unity in the Hodrick and Srivastava [107] specification. The results are somewhat inconsistent since McCurdy and Morgan's [154] specification fails to reject the null for the UK pound, the

TABLE XXVI
Coefficient estimates for daily futures data, 1974–1983 from McCurdy and Morgan [154]

Currency data	$\hat{\gamma}_0$	$\hat{\gamma}_1$	$\hat{\gamma}_2$	$\hat{\gamma}_3$	$\hat{\gamma}_4$	$\hat{\alpha}_0$	$\hat{\alpha}_1$	$\hat{\beta}_1$
BP	-0.0306 (0.0113)	0.015 (0.021)	0.084 (0.024)	-0.189 (0.058)	0.144 (0.023)	0.0060 (0.0014)	0.0105 (0.014)	0.886 (0.014)
BP H&S**	-0.0443 (0.0129)	0.924 (0.017)	0.078 (0.025)	-0.090 (0.055)	0.120 (0.025)	0.0059 (0.0013)	0.105 (0.014)	0.887 (0.013)
CD	-0.0108 (0.0048)	0.054 (0.025)				0.0003 (0.0002)	0.078 (0.012)	0.922 (0.012)
CD H&S	-0.0127 (0.0050)	0.931 (0.027)				0.0003 (0.002)	0.078 (0.013)	0.921 (0.013)
DM	-0.0310 (0.0165)	-0.044 (0.021)		-0.263 (0.070)	0.150 (0.035)	0.0148 (0.0033)	0.118 (0.015)	0.867 (0.015)
DM H&S	0.0202 (0.0204)	0.927 (0.022)		-0.289 (0.083)	0.143 (0.035)	0.0139 (0.0030)	0.119 (0.015)	0.867 (0.015)
JY	-0.0030 (0.0163)	-0.006 (0.026)	0.112 (0.043)	-0.127 (0.060)		0.0040 (0.0019)	0.064 (0.012)	0.930 (0.014)
JY H&S	0.0854 (0.0211)	0.878 (0.025)	0.069 (0.045)	-0.198 (0.074)		0.0035 (0.0017)	0.059 (0.011)	0.937 (0.011)
SF	-0.0159 (0.0134)	-0.008 (0.020)		-0.121 (0.060)	0.100 (0.029)	0.0032 (0.0012)	0.079 (0.012)	0.917 (0.012)
SF H&S	0.0253 (0.0175)	0.939 (0.019)		-0.114 (0.063)	0.102 (0.029)	0.0033 (0.0013)	0.079 (0.012)	0.917 (0.012)

Standard errors are shown in parenthesis.

* For the Canadian dollar and the Japanese yen the time period is 1977–1983.

** The H&S equations use the Hodrick and Srivastava [107] transformation of the futures prices, that is, $(F_t - S_{t-1})/S_{t-1}$ and $(F_{t-1} - S_{t-1})/S_{t-1}$ rather than $(\log F_t - \log F_{t-1})$ and $(\log F_{t-1} - \log F_{t-2})$ for the dependent variable and the regressor corresponding to $\hat{\gamma}_1$, respectively.

Japanese yen and the Swiss franc while the Hodrick and Srivastava [107] specification indicates rejection in all cases.

As noted above, McCurdy and Morgan [154, p. 15] like the logarithmic specification because they think the Hodrick and Srivastava [107] specification "is rejecting too frequently on account of a confounding influence of the spot price in the construction of the dependent variable and regressor." This intuition may be correct, although formally there is nothing wrong with either specification, but an alternative interpretation involves recognition that the power of the two tests may also differ. The argument is similar to the one involving autocorrelation of rates of depreciation of exchange rates. Using only the first differences of the logarithms of exchange rates one often cannot reject the hypothesis of no serial correlation, yet regressions of rates of depreciation on forward premiums do indicate some serial correlation. The signal-to-noise ratio in the regressors differs across the two specifications and that may account for the difference in inference.

McCurdy and Morgan [154] also are concerned with limit moves in the futures prices. In the early parts of the samples especially, the Chicago Mercantile Exchange had relatively small limits on daily price changes. This may induce some positive serial correlation when none would be present without the limits.

McCurdy and Morgan [154] take two actions to address the problem induced by limit moves. First, they conduct the daily analysis in the absence of data for which there occurred a limit move in the dependent variable. Second, they examine the hypothesis that the logarithm of the futures price is an unbiased predictor of the logarithm of the futures price in one week.

Since the inference without limit moves in daily data is not quantitatively different from the inference with limit moves, the weekly results are the interesting area for discussion. With this specification, only the Deutsche mark provides evidence against the null hypothesis. Since the dependent variable is the logarithmic change in the futures price and the regressor is the lagged dependent variable, this finding may be related to the discussion above regarding the power of alternative specifications that test the unbiasedness hypothesis. Reconciliation of these results with the results of Hodrick and Srivastava [107] indicating serial correlation in risk premiums appears to be an outstanding item on the research agenda.

7.3. Analysis of intraday data

In contrast to the conclusions of the preceding section, Feinstone [62] argues that intraday changes in exchange rates are serially uncorrelated. The analysis is conducted on each observation of trade for the September 1977 Deutsche mark futures contract as traded during July 1977.

Feinstone [62] examines the Q-statistic suggested by Ljung and Box [138] to test the null hypothesis of random serially uncorrelated changes in futures prices at either 30-second sampling intervals or 3-minute sampling intervals. The null hypothesis of no serial correlation is rejected at the 0.05 marginal level of significance only on 4 of 19 days for the 3-minute sample and only on 3 different days for the 30-second sample.

Feinstone [62] notes that instability in serial correlation coefficients could mask a potentially profitable trading rule. Hence, she examines two and three point (one ten-thousandth of a dollar) Alexander-style filter rules. The n-point trading rule requires the trader to buy the futures contract after an n-point rise in the futures price. When the futures price falls by n points from a subsequent peak, which could occur immediately, the trader sells the first contract and sells an additional futures contract to reverse the position. The smaller is n, the more transactions are generated and the higher are transactions costs.

The two and three point filter rules produced losses on eight and eleven of the nineteen trading days, respectively. The cumulative profits for the month were 42 points and −15 points, respectively. The profitability of the two-point filter rule is questionable on two grounds. First, it generated 264 round trips which is costly even for a local trader, and second, it presumes that the trader can execute at the price that triggers the filter.

Feinstone [62] examines the latter problem by analyzing a two-point trigger rule. The trade is presumed to be recorded at the price that triggered the filter rule if sixty seconds pass with no price change. Otherwise, the next higher price is used for a bid to buy and the next lower price for an offer to sell. Analysis of this trigger rule demonstrated losses on nine days and a cumulative loss for the month of 32 points.

Feinstone [62] also investigates the type of stochastic price process that characterizes intradaily movements in futures prices.

Since intradaily price processes appear leptokurtic, the Weiner process with its continuous increments of normally distributed price changes is rejected.[46] The alternative process that is investigated is the compound Poisson-normal process, or CPN.

The CPN process assumes that prices jump at the arrival of new information that occurs discretely at random times. The arrival of information is characterized by assuming that the number of jumps in an interval of length h is determined by an independent draw from a Poisson distribution. Given that a price jump occurs, the size of the price change is assumed to be normally distributed.

Examination of the CPN process indicates that it can produce leptokurtic price changes for some ranges of parameter values. Given the speed with which events are arriving and the size of the jump variance, Feinstone [62] derives the range of observation intervals over which price changes are more sharply peaked relative to the normal distribution.

The null hypothesis that a Poisson process characterizes the number of jumps can be rejected on only four of the nineteen days. These were also days for which white noise price changes were not supported by the data. Feinstone conjectures that intervention on these days by the Bundesbank may have been significant although no data are available to test this hypothesis.

Normality of price jumps appears to be supported by the data. Here care is taken to allow for the truncation of the normal induced by the discrete nature of price changes. Only price jumps in discrete points are allowed, hence any potential jump from −0.5 to 0.5 points will not be recorded. When the variance is allowed to be determined by the day's observations, normality of price jumps with a mean of zero is rejected at the 0.05 marginal level of significance only on one of the monthly trading days.

Pooling the sample one finds that it is easy to reject constancy of the parameter of the Poisson process across days of the month. Hence, a time-varying CPN process appears to be a reasonable

---

[46] Wasserfallen and Zimmerman [212] examine the intradaily price process for the spot exchange rate of Swiss francs per US dollar during nine separate days between 1978 and 1980. Distributions of relative changes over one minute intervals are highly leptokurtic, but they find that lengthening the observation interval to ten minutes produces convergence to a normal distribution. They also find no evidence of autocorrelation in the logarithmic changes in the spot rates.

characterization of the intradaily price process in the futures market for foreign currency. Such a process produces conditional heteroscedasticity which must be considered in estimation and testing of hypotheses. The nature of serial correlation in the presence of conditional heteroscedasticity is difficult to determine, and traditional tests such as the Q-statistic may be particularly affected by its presence.

It remains to be determined whether Feinstone's [62] serially uncorrelated process with no risk aversion is a better characterization of the data than the serially correlated processes motivated by risk aversion that are postulated by Hodrick and Srivastava [107].

8. CONCLUSIONS

This monograph has surveyed a rich set of empirical results that address questions regarding the efficiency of forward and futures markets for foreign exchange. In this section I will outline the central message of each of the preceding sections, and I will offer some suggestions for directions that the research may go.

Asset pricing theory was discussed only in a limited way in Section 2 since it was impossible to survey both the theoretical and the empirical literatures on this subject. The message of Section 2 was that it is relatively easy to develop rational expectations equilibrium models that allow for time-variation in the expected returns on assets. In such models the forward exchange rates or the futures prices of foreign currencies will generally not be unbiased expectations of future spot exchange rates. The bias also can move through time as the riskiness of assets denominated in different currencies fluctuates. The riskiness of an asset depends on the covariation of the return on the asset with the nominal intertemporal marginal rate of substitution. This, in turn, is a weighted average of the change in the purchasing power of a money where the weights depend on the real intertemporal marginal rate of substitution of consumption.

If inflation is stochastic, nominal assets are risky in real terms. Assets that have large real payoffs when the marginal utility of consumption is high are like insurance policies; they payoff in bad times. Consequently, risk averse investors willingly hold these assets, and in equilibrium they price them to have low expected returns; just like they willingly pay up front for insurance against bad states of the world.

Several challenges exist for the representative agent model of asset pricing. For instance, Mehra and Prescott [161] argue that the model cannot be consistent with the unconditional premium of equity returns over treasury security returns. Additionally, Grossman and Shiller [91] and Shiller [194] suggest that the model is incapable of reconciling the volatility of stock prices relative to the volatility of dividends even if discount rates are allowed to vary. Such results have been challenged by Kleidon [119, 120] and Marsh and Merton [150]. The ongoing nature of this debate will surely continue to have its international counterpart. The volatility of exchange rates is every bit as much of a puzzle as the volatility of aggregate stock prices.

Section 3 considered two sets of empirical tests that sought to determine the efficiency of the spot and forward markets for foreign exchange. The first were examinations of the time series properties and the distributions of exchange rates. The second were considerations of the hypothesis that the forward rate is an unbiased predictor of the future spot rate. Much confusion has been generated by claims that the exchange rate ought to follow a random walk in an efficient market. This is simply false. There can be an expected rate of depreciation of one currency relative to another, and hence tests of the autocorrelations of exchange rates are relatively uninteresting except in the sense of providing interesting stylized facts. The statistical time series analysis indicates that exchange rates are so volatile that it is difficult to distinguish them from random walks. A potential problem with such studies is that they typically assume that the conditional variance of exchange rates is constant.

Recent works by Boothe and Glassman [20] and Hsieh [112] discussed in Section 3 suggest that the distribution of exchange rates may not be constant through time. Consequently, Section 3 and other sections addressed studies of the unbiasedness hypothesis that assume conditional homoscedasticity and ones that allow for certain forms of conditional heteroscedasticity. The Generalized Method of Moments estimators developed by Hansen [96] may have an advantage in this regard over the maximum likelihood techniques that assume conditional homoscedasticity. Both techniques are discussed in Section 3.

Evidence against the unbiasedness hypothesis is not concentrated just in Section 3. The evidence appears to be very strong and consistent across currencies, maturities and time periods.

Section 4 discusses several interpretations of the rejection of the unbiasedness hypothesis. Consideration is given to several viewpoints. The nature of the tests of unbiasedness is that they rely on asymptotic distribution theory to generate distributions of test statistics, and empirical researchers are forced to assume that the data satisfy an ergodicity assumption. One possibility is that the small sample distributions of the test statistics simply do not coincide with the asymptotic theoretical distributions. The work of Korajczyk [122] and the interesting simulations of tests of asset pricing models conducted by Mattey and Meese [151] suggest that this may indeed be a real possibility.

A second line of criticism of the typical tests in this area concerns the validity of the ergodicity assumption. It is relatively easy to envision scenarios that lead to failure of the ergodicity assumption. Whenever there are potential changes in government policy processes that have not occurred in the sample, the data are not ergodic. Ergodicity is also a problem if there are events that occur during the sample but not with the appropriate frequency to correspond to their a priori probability. This is the classic 'Peso Problem' of too few devaluations during a fixed rate regime. Lizondo [137] demonstrates how prospects of a devaluation that does not occur during a sample can distort inference, and Obstfeld [170] provides a nice example under flexible exchange rates of the incorrect inference that arises if agents are rationally expecting an event that does not occur during the sample. Such problems plague all modern time series methods that rely on an assumption of ergodicity. No one has been able to offer a suggestion of how to deal with this problem short of modeling the expected events that concern individuals that are not occurring with sufficient frequency.

Explicit examples of the failure of the rational expectations assumption because of learning by agents have also started to be produced. Lewis [136] and Stulz [205] question the assumption of rational expectations and demonstrate how serially correlated forecast errors could result if agents are learning about a government policy. These caveats must always be kept in mind in interpreting the results of any study employing the rational expectations econometric methodology.

Much of the discussion in Section 4 concerned Fama's [59] interesting decomposition argument that has generated considerable controversy and subsequent analysis. Fama demonstrated that the nature of the rejection of the unbiasedness hypothesis (if the statistics are taken as correct) implies that the risk premiums and expected rates of depreciation covary negatively and that variability of risk premiums is greater than the variability of expected rates of depreciation. Although Fama [59] found such results troublesome and suggested that they might represent evidence against an efficient market, Hodrick and Srivastava [108] provided an analysis which showed that such results need not be taken as inconsistent with the theory presented in Section 2. The outstanding issue appears to be what would be the source of the volatility.

The evidence in Giovannini and Jorion [86] suggests that movement in the conditional risk premiums of the return on the US stock market covary with the conditional risk premiums in the forward market. Hence, common sources of riskiness that are traced to government policies and technological change may be an explanation.

The investigation of forecasts of exchange rates by Froot and Frankel [79], discussed in Section 6, suggests that problems with the data, such as are caused by learning on the part of agents about the international environment may be more important than previous researchers have thought.

Another interpretation of the rejection of the unbiasedness hypothesis discussed in Section 4 is that the data indicate an inefficiency in the foreign exchange market. The claims of excessive trading rule profitability are explored, and the findings of some studies are examined statistically. Unfortunately, without an unrejected model of expected returns that vary through time, it is difficult to know whether the apparent profitability of some of the trading strategies is simply consistent with changes in the riskiness of currencies or whether the evidence is truly a market inefficiency. Reconciliation of the filter rule studies with the models of time varying risk premiums is a challenging area of future work.

Section 5 explores models that have been examined empirically that could potentially reconcile the rejection of the unbiasedness hypothesis with market efficiency. Models that use only financial data and models that use other variables that are fundamental to the markets are discussed. None of the current models appears capable of the task, but some of the results, especially those of Korajczyk [122] and Mark [148], appear to hold out promise that a model may be able to be developed that is not inconsistent with the data.

Section 7 explores research in the futures pricing area. The availability of daily data in this area is only beginning to be explored. Initial empirical results indicate that even though theory does not require that futures prices equal forward rates, there is no discernible difference between them. Hence, results on the unbiasedness hypothesis carry over to futures prices, too. This lead Hodrick and Srivastava [107] to examine whether the futures price on a given day is an unbiased predictor of the futures price on the following day. They concluded that it is not although their results have been questioned by McCurdy and Morgan [154].

One problem that arises in the empirical work that is discussed in Section 7 is generated by the nature of the contract in the futures markets. The forecast horizon of the futures prices evolves continually since the maturity date of the contract is fixed as the third Wednesday in March, June, September and December. Hence, inference must be based on large sample theory that considers a hypothetical vector of futures prices that is sampled in a particular way. Initial work in this area appears to be generating results that are somewhat contradictory as regards the nature of time variation in risk premiums and whether daily variation is consistent with the monthly variability implied by forward rates.

Understanding the risk-return tradeoff in international financial markets is an important issue to many people and organizations, both public and private, throughout the world. The studies discussed in this monograph form a foundation on which future work can be based. We do not yet have a model of expected returns that fits the data. International finance is no worse off in this respect than more traditional areas of finance.

Asset pricing in general is also an exciting area of research given recent empirical results that have overturned simple models such as the static capital asset pricing model that had prevailed for more than a decade. One issue that needs to be addressed more formally is the role of learning in determining interest rates, exchange rates and stock prices. Another difficult area of research is quantifying the influence of government policies on asset prices. The rational expectations revolution in macroeconomics alerted us to the role of the influence of expected future policies on current variables. Is history useful in determining the expected future path of government policies or must we bring more theory to bear on the problem? My intuition here is that simple extrapolations of the past (or autoregressive time series models) are not very useful in determining expectations of future government policies.

The world is rapidly changing and becoming more integrated. A global marketplace in assets and commodities is emerging as technological change has decreased the cost of communication around the world. Will governments embrace this opportunity for increased intersectoral and intertemporal trade or will they shy away and attempt to erect protectionist barriers that prohibit the flows of goods and the flows of wealth? What causes exchange rate volatility and is it harmful to trade and welfare? Studies that seek answers to these and other issues in international finance and macroeconomics must base their models on some intertemporal paradigm. The work presented here suggests that simple models may not work well, but we have only begun to develop the first models based on rational maximizing agents.

As economists, we would like to be able to advise policy makers on the conduct of policy. We would like to be able to advise firms on the nature of exchange risk. We need to know how well international financial markets are functioning in the allocation of savings and investment throughout the world. We need to know how monetary and fiscal policies and trade policies ought to be conducted. Answers to these questions are important, and I hope that the readers of this monograph will be able to use the research presented here in their quest for the truth.

References

[1] Adams, F. C. and R. S. Boyer: "Forward Premia and Risk Premia in a Simple Model of Exchange Rate Determination," Working Paper, University of Western Ontario, June 1986.

[2] Adler, M. and B. Dumas: "International Portfolio Choice and Corporation Finance: A Survey," *Journal of Finance*, **38** (1983), 925–984.

[3] Adler, M. and B. Lehmann: "Deviations from Purchasing Power Parity in the Long Run," *Journal of Finance*, **38** (1983), 1471–1487.

[4] Agmon, T. and Y. Amihud: "The Forward Exchange Rate and the Prediction of the Future Spot Rate," *Journal of Banking and Finance*, **5** (1981), 425–437.

[5] Akaike, H.: "Information Theory and an Extension of the Maximum Likelihood Principle," in *2nd International Symposium on Information Theory*, ed. B. N. Petrov and F. Csaki,. Budapest: Akademiai Kiado, 1973.

[6] Alexander, S. S.: "Price Movements in Speculative Markets: Trends or Random Walks," *Industrial Management Review*, **2** (1961), 7–26.

[7] Aliber, R. Z.: "The Interest Rate Parity Theorem: A Reinterpretation," *Journal of Political Economy*, **81** (1973), 1451–1459.

[8] Aliber, R. Z.: "Attributes of National Monies and the Independence of National Monetary Policies," in *National Monetary Policies and the International Financial System*, ed. by R. Z. Aliber, Chicago: University of Chicago Press, 1974.

[9] Arrow, K.: "The Role of Securities in the Optimal Allocation of Risk-Bearing," *Review of Economic Studies*, **31** (1984), 91–96.

[10] Bailey, R. W., R. T. Baillie, and P. C. McMahon: "Interpreting Econometric Evidence on Efficiency in the Foreign Exchange Market," *Oxford Economic Papers*, **36** (1984), 67–85.

[11] Baillie, R. T.: "Prediction from the Dynamic Simultaneous Equation Model with Vector Autoregressive Errors," *Econometrica*, **49** (1981), 1331–1337.

[12] Baillie, R. T., R. E. Lippens, and P. C. McMahon: "Testing Rational Expectations and Efficiency in the Foreign Exchange Market," *Econometrica*, **51** (1983), 553–564.

[13] Bilson, J. F. O.: "The Speculative Efficiency Hypothesis," National Bureau of Economic Research Working Paper No. 747, 1980.

[14] Bilson, J. F. O.: "The 'Speculative Efficiency' Hypothesis," *Journal of Business*, **54** (1981), 435–452.

[15] Bilson, J. F. O. and D. Hsieh: "The Risk and Return of Currency Speculation," mimeo, University of Chicago, 1984.

[16] Black, F.: "The Pricing of Commodity Contracts," *Journal of Financial Economics*, **3** (1976), 167–179.

[17] Bollerslev, T.: "Generalized Autoregressive Conditional Heteroscedasticity," *Journal of Econometrics*, forthcoming (1987).

[18] Boothe, P.: *Estimating the Structure and Efficiency of the Canadian Foreign Exchange Market: 1971–1978.* University of British Colombia Ph.D. Thesis, May 1981.

[19] Boothe, P., K. Clinton, A. Cote, and D. Longworth: *International Asset Substitutability: Theory and Evidence for Canada.* Ottawa, Canada: Bank of Canada, 1985.

[20] Boothe, P. and D. Glassman: "The Statistical Distribution of Exchange Rates: Empirical Evidence and Economic Implications," Research Paper No. 85–22, Department of Economics, University of Alberta, 1985.

[21] Boothe, P. and D. Longworth: "Foreign Exchange Market Efficiency Tests: Implications of Recent Findings," *Journal of International Money and Finance*, **5** (1986), 135–152.

[22] Boothe, P. and C. Sawchuk: "The Peso Problem and Unexploited Profits in the Canadian Foreign Exchange Market," Research Paper No. 85–21, Department of Economics, University of Alberta, 1985.

160 R. J. HODRICK

[23] Boyer, R.: "The Relation Between the Forward Exchange Rate and the Expected Future Spot Rate," *Intermountain Economic Review,* **8** (1977), 14–21.
[24] Branson, W. H. and D. W. Henderson: "The Specification and the Influence of Asset Markets," in *Handbook of International Economics: Volume 2* ed. by R. W. Jones and P. B. Kenen. Amsterdam: North-Holland, 1985.
[25] Breeden, D.: "An Intertemporal Asset Pricing Model with Stochastic Consumption and Investment Opportunities," *Journal of Financial Economics,* **7** (1979), 265–296.
[26] Brock, W. A.: "Asset Pricing in an Economy with Production: 'Selective' Survey of Recent Work on Asset-Pricing Models," in *Dynamic Optimization and Mathematical Economics,* ed. By Pan-Tai Lin. New York: Alenum Press, 1980.
[27] Brown, B. W. and S. Maital: "What Do Economists Know? An Empirical Study of Experts' Expectations," *Econometrica,* **49** (1981), 491–504.
[28] Brown, R. L., J. Durbin, and J. M. Evans: "Techniques for Testing the Constancy of Regression Relationships Over Time," *Journal of the Royal Statistical Society,* **37** (1975), 149–182.
[29] Campbell, J. Y. and R. H. Clarida: "The Term Structure of Euromarket Interest Rates: An Empirical Investigation," *Journal of Monetary Economics,* **19** (1987), 25–44.
[30] Chow, G.: "Tests of Equality Between Subsets of Coefficients in Two Linear Regressions," *Econometrica,* **28** (1960), 591–605.
[31] Cornell, B.: "Spot Rates, Forward Rates and Exchange Market Efficiency," *Journal of Financial Economics,* **5** (1980), 55–65.
[32] Cornell, B. and J. K. Dietrich: "The Efficiency of the Market for Foreign Exchange Under Floating Exchange Rates," *Review of Economics and Statistics,* (1978), 111–120.
[33] Cornell, B. and M. Reinganum: "Forward and Futures Prices: Evidence from the Foreign Exchange Markets," *Journal of Finance,* **36** (1981), 1035–1045.
[34] Cox, J., J. Ingersoll, and S. Ross: "The Relation Between Forward Prices and Futures Prices," *Journal of Financial Economics,* **9** (1981), 321–346.
[35] Cumby, R., J. Huizinga, and M. Obstfeld: "Two-Step Two-Stage Least Squares Estimation in Models with Rational Expectations," *Journal of Econometrics,* **21** (1983), 333–355.
[36] Cumby, R. E. and M. Obstfeld: "A Note on Exchange-Rate Expectations and Nominal Interest Differentials: A Test of the Fisher Hypothesis," *Journal of Finance,* **36** (1981), 697–704.
[37] Cumby, R. E. and M. Obstfeld: "International Interest-Rate and Price-Level Linkages Under Flexible Exchange Rates: A Review of Recent Evidence," in *Exchange Rates: Theory and Practice* ed. by J. F. O. Bilson and R. Marston. Chicago: University of Chicago Press for the National Bureau of Economic Research, 1984.
[38] Danker, D. J., R. A. Haas, D. W. Henderson, S. A. Symansky, and R. W. Tryon: "Small Empirical Models of Exchange Market Intervention: Applications to Germany, Japan, and Canada," Staff Studies No. 135, Board of Governors of the Federal Reserve System, Washington, D.C. 1985.
[39] Debreu, G.: *Theory of Value.* New Haven: Yale University Press, 1959.
[40] Dhyrmes, P.: *Econometrics: Statistical Foundations and Applications.* New York: Springer-Verlag, 1974.
[41] Dickey, D. A. and W. A. Fuller: "Distribution of the Estmators in

Autoregressive Time Series with a Unit Root," *Journal of the American Statistical Association*, **74** (1979), 427–431.

[42] Domowitz, I. and C. Hakkio: "Conditional Variance and the Risk Premium in the Foreign Exchange Market," *Journal of International Economics*, **19** (1985), 47–66.

[43] Dooley, M. P. and P. Isard: "Capital Controls, Political Risk, and Deviations from Interest Rate Parity," *Journal of Political Economy*, **88** (1980), 370–84.

[44] Dooley, M. P. and J. Shafer: "Analysis of Short-Run Exchange Rate Behavior: March, 1973 to September, 1975," International Finance Discussion Paper No. 123, Federal Reserve Board, Washington, 1976.

[45] Dooley, M. P. and J. Shafer: "Analysis of Short-Run Exchange Rate Behavior: March, 1973 to November 1981," in *Exchange Rate and Trade Instability: Causes, Consequences, and Remedies*, ed. by D. Bigman and T. Taya. Cambridge, MA: Ballinger, 1983.

[46] Dornbusch, R.: "Exchange Rate Economics: Where Do We Stand?" *Brookings Papers on Economic Activity*, 1980 (1), 143–194.

[47] Dornbusch, R.: "Exchange Risk and the Macroeconomics of Exchange Rate Determination," in *The Internationalization of Financial Markets and National Economic Policy*, ed. by R. Hawkins, R. Levich, and C. Wihlborg. Greenwich, Conn.: JAI Press, 1982.

[48] Dornbusch, R.: "Equilibrium and Disequilibrium Exchange Rates," *Zeitschift fur Wirtschafts und Sozialwissenshaften*, **102** (1982), 573–599.

[49] Dunn, K., and K. Singleton: "Modeling the Term Structure of Interest Rates Under Non-separable Utility and Durability of Goods," *Journal of Financial Economics*, **17** (1986), 27–56.

[50] Dusak, K.: "Futures Trading and Investor Returns: An Investigation of Commodity Market Risk Premiums," *Journal of Political Economy*, **81** (1973), 1387–1406.

[51] Edwards, S.: "Exchange Rates and 'News': A Multi-Currency Approach," *Journal of International Money and Finance*, **1** (1982), 211–224.

[52] Efron, B.: *The Jackknife, the Bootstrap, and Other Resampling Plans*. Philadelphia: Society of Industrial and Applied Mathematics, 1982.

[53] Eichenbaum, M. and L. P. Hansen: "Uncertainty, Aggregation, and the Dynamic Demand for Consumption Goods," mimeo, Carnegie-Mellon University, 1984.

[54] Engel, C. H.: "Testing for the Absence of Expected Real Profits from Forward Market Speculation," *Journal of International Economics*, **17** (1984), 309–324.

[55] Engle, R.: "Autoregressive Conditional Heteroscedasticity with Estimates of the Variance of the United Kingdom Rate of Inflation," *Econometrica*, **50** (1982), 987–1007.

[56] Engle, R. F., D. M. Lilien, and R. P. Robins: "Uncertainty and the Term Structure," University of California, San Diego Discussion Paper No. 82–84.

[57] Fama, E.: "Efficient Capital Markets: A Review of Theory and Empirical Work," *Journal of Finance*, **25** (1970), 383–417.

[58] Fama, E.: *Foundations of Finance*. New York: Basic Books, 1976.

[59] Fama, E.: "Forward and Spot Exchange Rates," *Journal of Monetary Economics*, **14** (1984), 319–38.

[60] Fama, E. and A. Farber: "Money, Bonds and Foreign Exchange," *American Economic Review*, **69** (1979), 269–282.

[61] Farber, A., R. Roll, and B. Solnik: "An Empirical Study of Risk Under Fixed

162   R. J. HODRICK

and Flexible Exchange Rates," in *Stabilization of the Domestic and International Economy* ed. by K. Brunner and A. H. Meltzer, Vol. 5 Carnegie-Rochester Conference Series on Public Policy, supplement to the *Journal of Monetary Economics,* (1977), 235–265.

[62] Feinstone, L.: "Intradaily Market Efficiency and Price Processes in the Futures Market in Foreign Exchange," mimeo, University of Rochester, 1985.

[63] Flood, R. P.: "Explanations of Exchange-Rate Volatility and Other Empirical Regularities in Some Popular Models of the Foreign Exchange Market," in *The Costs and Consequences of Inflation,* ed. by K. Brunner and A. H. Meltzer, Vol. 15 Carnegie-Rochester Conference Series on Public Policy, supplement to the *Journal of Monetary Economics,* (1981), 219–250.

[64] Flood, R. P. and P. M. Garber: "A Model of Stochastic Process Switching," *Econometrica,* **51** (1983), 537–552.

[65] Frankel, J.: "The Diversifiability of Exchange Risk," *Journal of International Economics,* **9** (1979), 379–393.

[66] Frankel, J.: "Tests of Rational Expectations in the Forward Exchange Market," *Southern Economic Journal,* **46** (1980), 1083–1101.

[67] Frankel, J.: "In Search of the Exchange Risk Premium: A Six-Currency Test Assuming Mean Variance Optimization," *Journal of International Money and Finance,* **1** (1982), 255–274.

[68] Frankel, J.: "The Implications of Mean-Variance Optimization for Four Questions in International Macroeconomics," *Journal of International Money and Finance,* **5** (1986), S53–S76.

[69] Frankel, J. and C. Engel: "Do Asset Demands Optimize over the Mean and Variance of Real Returns? A Six-Currency Test," *Journal of International Economics,* **17** (1984), 309–23.

[70] Frankel, J. A. and K. Froot: "Using Survey Data to Test Some Standard Propositions Regarding Exchange Rate Expectations," National Bureau of Economic Research Working Paper No. 1672, 1985.

[71] Frenkel, J. A.: "The Forward Exchange Rate, Expectations, and the Demand for Money: The German Hyperinflation," *American Economic Review,* **67** (1977), 653–670.

[72] Frenkel, J. A.: "Flexible Exchange Rates, Prices, and the Role of 'News': Lessons from the 1970s," *Journal of Political Economy,* **89** (1981), 655–705.

[73] Frenkel, J. A. and R. Levich: "Covered Interest Arbitrage: Unexploited Profits?" *Journal of Political Economy,* **83** (1975), 325–338.

[74] Frenkel, J. A. and R. Levich: "Transaction Costs and Interest Arbitrage: Tranquil versus Turbulent Periods," *Journal of Policical Economy,* **86** (1977), 1209–1226.

[75] Frenkel, J. A. and M. Mussa: "The Efficiency of the Foreign Exchange Market and Measures of Turbulence," *American Economic Review,* **70** (1980), 374–81.

[76] Frenkel, J. A. and A. Razin: "Stochastic Prices and Tests of Efficiency of Foreign Exchange Markets," *Economic Letters,* **6** (1980), 165–170.

[77] Friedman, D. and S. Vandersteel: "Short-Run Fluctuations in Foreign Exchange Rates: Evidence from the Data, 1973–79," *Journal of International Economics,* **13** (1982), 171–186.

[78] Friedman, M.: *Essays in Positive Economics.* Chicago: University of Chicago Press, 1953.

[79] Froot, K. A. and J. A. Frankel: "Interpreting Tests of Forward Discount Bias Using Survey Data on Exchange Rate Expectations," mimeo, Sloan School of Management, Massachusetts Institute of Technology, 1986.

[80] Fuller, W. A.: *Introduction to Statistical Time Series*. New York: John Wiley and Sons, 1976.
[81] Garber, P. and R. King: "Deep Structural Excavation? A Critique of Euler Equation Methods," N.B.E.R. Technical Paper No. 31, 1983.
[82] Geweke, J. and E. Feige: "Some Joint Tests of the Efficiency of Markets for Forward Foreign Exchange," *Review of Economic Statistics*, **61** (1979), 334–341.
[83] Gibbons, M.: "Multivariate Tests of Financial Models: A New Approach," *Journal of Financial Economics*, **10** (1982), 3–27.
[84] Gibbons, M. and W. Ferson: "Testing Asset Pricing Models with Changing Expectations and an Unobservable Market Portfolio," *Journal of Financial Economics*, **14** (1985), 217–236.
[85] Giddy, I. H. and G. Dufey: "The Random Behavior of Flexible Exchange Rates," *Journal of International Business Studies*, **6** (1975), 1–32.
[86] Giovannini, A. and P. Jorion: "Interest Rates and Risk Premia in the Stock Market and in the Foreign Exchange Markets," *Journal of International Money and Finance*, **6** (1987).
[87] Goodman, S. and R. Jaycobs: "Double Up and Prosper," *Euromoney*, August (1983), 132–139.
[88] Grauer, F., R. Litzenberger, and R. Stehle: "Sharing Rules and Equilibrium in an International Capital Market Under Uncertainty," *Journal of Financial Economics*, **3** (1976), 233–256.
[89] Gregory, A. W. and T. H. McCurdy: "Testing the Unbiasedness Hypothesis in the Forward Foreign Exchange Market: A Specification Analysis," *Journal of International Money and Finance*, **3** (1984), 357–368.
[90] Grenander, U.: "On the Estimation of Regression Coefficients in the Case of an Autorcorrelated Disturbance," *Annals of Mathematical Statistics*, **25** (1954), 252–272.
[91] Grossman, S. and R. J. Shiller: "The Determinants of the Variability of Stock Market Prices," *American Economic Review*, **71** (1981), 222–227.
[92] Hakkio, C. S.: "The Term Structure of the Forward Premium," *Journal of Monetary Economics*, **8** (1981), 41–58.
[93] Hakkio, C. S.: "Expectations and the Forward Exchange Rate," *"International Economic Review*, **22** (1981), 663–678.
[94] Hakkio, C. S.: "Discussion of Risk Averse Speculation in the Forward Foreign Exchange Market: An Econometric Analysis of Linear Models," in *Exchange Rates and International Macroeconomics*, ed. by J. A. Frenkel. Chicago: University of Chicago Press for National Bureau of Economic Research, 1983.
[95] Hannan, E.: *Multiple Time Series*. New York: John Wiley and Sons, 1970.
[96] Hansen, L. P.: "Large Sample Properties of Generalized Method of Moments Estimators," *Econometrica*, **50** (1982), 1029–1054.
[97] Hansen, L. P. and R. J. Hodrick: "Forward Exchange Rates as Optimal Predictors of Future Spot Rates: An Econometric Analysis," *Journal of Political Economy*, **88** (1980), 829–853.
[98] Hansen, L. P. and R. J. Hodrick: "Risk Averse Speculation in the Forward Foreign Exchange Market: An Econometric Analysis of Linear Models," in *Exchange Rates and International Macroeconomics*, ed. by J. A. Frenkel. Chicago: University of Chicago Press for National Bureau of Economic Research, 1983.
[99] Hansen, L. P. and S. Richard: "A General Approach for Deducing Testable Restrictions Implied by Asset Pricing Models," mimeo, Carnegie-Mellon University, 1984.

164 R. J. HODRICK

[100] Hansen, L. P. and K. J. Singleton: "Generalized Instrumental Variables Estimation of Nonlinear Rational Expectations Models," *Econometrica*, **50** (1982), 1269–1286.
[101] Hansen, L. P. and K. J. Singleton: "Consumption, Risk Aversion and the Temporal Behavior of Asset Returns," *Journal of Political Economy*, **91** (1983), 249–265.
[102] Hansen, L. P. and K. J. Singleton: "Errata," *Econometrica*, **52** (1984), 267–268.
[103] Hasza, D. P. and W. A. Fuller, "Estimation of Autoregressive Processes with Unit Roots," *Annals of Statistics*, **7** (1979), 1106–1120.
[104] Henderson, D. W.: "Exchange Market Intervention Operations: Their Role in Financial Policy and their Effects," in *Exchange Rate Theory and Practice*, edited by J. F. O. Bilson and R. Marston. Chicago: University of Chicago Press for the National Bureau of Economic Research, 1984.
[105] Hodrick, R. J.: "International Asset Pricing with Time-Varying Risk Premia," *Journal of International Economics*, **11** (1981), 573–587.
[106] Hodrick, R. J. and S. Srivastava: "An Investigation of Risk and Return in Forward Foreign Exchange," *Journal of International Money and Finance*, **3** (1984), 1–29.
[107] Hodrick, R. J. and S. Srivastava: "Foreign Currency Futures," National Bureau of Economic Research Working Paper No. 1743, (1985), forthcoming *Journal of International Economics*.
[108] Hodrick, R. J. and S. Srivastava: "The Covariation of Risk Premiums and Expected Future Spot Exchange Rates," *Journal of International Money and Finance*, **5** (1986), S5–S22.
[109] Hosking, J. R. M.: "The Multivariate Portmanteau Statistic," *Journal of the American Statistical Association*, **75** (1980), 602–608.
[110] Hsieh, D.: "A Heteroscedasticity-Consistent Covariance Estimator for Time Series Regressions," *Journal of Econometrics*, **22** (1983), 281–290.
[111] Hsieh, D.: "Tests of Rational Expectations and No Risk Premium in Forward Exchange Markets," *Journal of International Economics*, **17** (1984), 173–184.
[112] Hsieh, D.: "The Statistical Properties of Daily Foreign Exchange Rates: 1974–1983, "mimeo, University of Chicago, 1985.
[113] Huang, R. D.: "Some Alternative Tests of Forward Rates as Predictors of Future Spot Rates," *Journal of International Money and Finance*, **3** (1984), 153–168.
[114] Ibbotson, R. G. and R. A. Sinquefield: "Stocks, Bonds, Bills and Inflation: Year-by-year Historial Returns, 1926–1974," *Journal of Business*, **49** (1976), 11–47.
[115] Ibbotson, R. G. and R. A. Sinquefield: *Stocks, Bonds, Bills and Inflation: The Past and the Future*, Financial Analysts Research Foundation, Charlottesville, VA, 1982.
[116] Jagannathan, R.: "Three Essays on the Pricing of Derivative Claims," Ph.D. Dissertation, Carnegie-Mellon University, 1983.
[117] Jagannathan, R.: "An Investigation of Commodity Futures Pricing Using the Consumption-Based Intertemporal Capital Asset Pricing Model," *Journal of Finance*, **40** (1985), 175–191.
[118] Jones, R. W. and P. B. Kenen: *Handbook of International Economics: Vol. 2.* Amsterdam: North-Holland, 1985.
[119] Kleidon, A.: "Variance Bounds Tests and Stock Price Valuation Models," *Journal of Political Economy*, **94** (1986), 953–1001.

[120] Kleidon, A.: "Bias in Small Sample Tests of Stock Price Rationality," *Journal of Business*, **59** (1976), 237–262.

[121] Kohlhagen, S. W.: *The Behavior of Foreign Exchange Markets: A Critical Survey of the Empirical Literature*, New York University Monograph Series in Finance and Economics No. 3, Salomon Brothers Center, 1978.

[122] Korajczyk, R. A.: "The Pricing of Forward Contracts for Foreign Exchange," *Journal of Political Economy*, **93** (1985), 346–368.

[123] Kouri, P.: "International Investment and Interest Rate Linkages Under Flexible Exchange Rates," in *The Political Economy of Monetary Reform*, ed. by R. Z. Aliber. London: Macmillan, 1977.

[124] Krasker, W. S.: "The 'Peso Problem' in Testing the Efficiency of Forward Exchange Markets," *Journal of Monetary Economics*, **6** (1980), 276–296.

[125] Krugman, P. R.: "Purchasing Power Parity and Exchange Rates," *Journal of International Economics*, **8** (1978) 397–407.

[126] Krugman, P. R.: "Consumption Preferences, Asset Demands, and Distribution Effects in International Financial Markets," N.B.E.R. Working Paper No. 651, 1981.

[127] Leamer, E. E.: *Specification Searches: Ad Hoc Inference with Nonexperimental Data*. New York: John Wiley and Sons, 1978.

[128] Levich, R.: "Further Results on the Efficiency of Markets for Foreign Exchange," in *Managed Exchange-Rate Flexibility: The Recent Experience*, Federal Reserve Bank of Boston Conferences Series No. 20. Boston: Federal Reserve Bank of Boston, 1978.

[129] Levich, R.: "On the Efficiency of Markets for Foreign Exchange," in *International Economic Policy: Theory and Evidence*, ed. by R. Dornbusch and J. A. Frenkel. Baltimore: Johns Hopkins University Press, 1979.

[130] Levich, R.: "Analyzing the Accuracy of Foreign Exchange Forecasting Services: Theory and Evidence," National Bureau of Economic Research Working Paper No. 336, 1979.

[131] Levich, R.: "Evaluating the Performance of the Forecasters," in *The Management of Foreign Exchange Risk, Second Edition*, ed. by B. Antl and R. Ensor, London: Euromoney Publications, 1982.

[132] Levich, R.: "Currency Forecasters Lose Their Way, "*Euromoney*, (1983), 140–147.

[133] Levich, R.: "Empirical Studies of Exchange Rates: Price Behavior, Rate Determination and Market Efficiency," in *Handbook of International Economics: Volume 2* ed. by R. W. Jones and P. B. Kenen. Amsterdam: North-Holland, 1985.

[134] Levy, E. and A. R. Nobay: "The Speculative Efficiency Hypothesis: A Bivariate Analysis," *Economic Journal*, (1986).

[135] Lewis, K. K.: "Risk Aversion and the Effectiveness of Sterilized Intervention," mimeo, New York University, 1986.

[136] Lewis, K. K.: "The Implications of Stochastic Policy Processes for the 'Peso Problem' in Flexible Exchange Rates and Other Prices," mimeo, New York University, 1986.

[137] Lizondo, J. S.: "Foreign Exchange Futures Prices Under Fixed Exchange Rates," *Journal of International Economics*, **14** (1983), 69–84.

[138] Ljung, G. M. and G. E. P. Box: "Studies in the Modeling of Discrete Time Series: 3. A Modification of the Overall χ^2 Test for Lack of Fit in Time Series Models," Technical Report No. 477, Department of Statistics, University of Wisconsin, 1976.

[139] Longworth, D.: "Testing the Efficiency of the Canadian-U.S. Exchange Market Under the Assumption of No-Risk Premium," *Journal of Finance*, **36** (1981), 43–49.
[140] Longworth, D., P. Boothe, and K. Clinton: *A Study of the Efficiency of the Foreign Exchange Market*. Ottawa, Canada: Bank of Canada, 1983.
[141] Loopesko, B. E.: "Relationships among Exchange Rates, Intervention, and Interest Rates: An Empirical Investigation," *Journal of International Money and Finance*, **3** (1984), 257–278.
[142] Lucas, R. E. Jr.: "Econometric Policy Evaluation: A Critique," in *The Phillips Curve and Labor Markets*, ed. by K. Brunner and A. H. Meltzer, Vol. 1 of the Carnegie-Rochester Conference Series on Public Policy, supplement to the *Journal of Monetary Economics*, (1976), 19–46.
[143] Lucas, R. E. Jr.: "Asset Pricing in an Exchange Economy," *Econometrica*, **46** (1978), 1429–1445.
[144] Lucas, R. E. Jr.: "Equilibrium in a Pure Currency Economy," *Economic Inquiry*, **18** (1980), 203–220.
[145] Lucas, R. E. Jr.: "Interest Rates and Currency Prices in a Two-Country , World," *Journal of Monetary Economics*, **10** (1982), 335–360.
[146] Lucas, R. E. Jr.: "Money in a Theory of Finance," in *Essays on Macroeconomic Implications of Financial and Labor Markets and Political Processes*, ed. by K. Brunner and A. H. Meltzer, Vol. 21, Carnegie-Rochester Conference Series on Public Policy, supplement to the *Journal of Monetary Economics*, (1984), 9–45.
[147] Lucas, R. E. Jr. and N. L. Stokey: "Optimal Fiscal and Monetary Policy in an Economy Without Capital," *Journal of Monetary Economics*, **12** (1983), 55–93.
[148] Mark, N. C.: "On Time Varying Risk Premia in the Foreign Exchange Market: An Econometric Analysis," *Journal of Monetary Economics*, **16** (1985), 3–18.
[149] Mark, N. C.: "Some Evidence on the International Inequality of Real Interest Rates," *Journal of International Money and Finance*, **4** (1985), 189–208.
[150] Marsh, T. A. and R. C. Merton: "Dividend Variability and Variance Bounds Tests for the Rationality of Stock Market Prices," *American Economic Review*, **76** (1986), 483–498.
[151] Mattey, J. and R. Meese: "Empirical Assessment of Present Value Relations," *Econometric Reviews*, (1987).
[152] McCormick, F.: "Covered Interest Arbitrage: Unexploited Profits? Comment," *Journal of Political Economy*, **87** (1979), 411–417.
[153] McCulloch, J. H.: "Operational Aspects of the Siegel Paradox: Comment," *Quarterly Journal of Economics*, **89** (1975), 170–174.
[154] McCurdy, T. H. and I. G. Morgan: "Tests of the Martingale Hypothesis for Foreign Currency Futures With Time Varying Volatility," Working Paper, Queen's University, July 1986.
[155] McFarland, J. W., R. R. Pettit, and S. K. Sung: "The Distribution of Foreign Exchange Price Changes: Trading Day Effects and Risk Measurement," *Journal of Finance*, **37** (1982), 693–715.
[156] McKinnon, R.: *Money in International Exchange: The Convertible Currency System*. New York: Oxford University Press, 1979.
[157] Meese, R.: "Empirical Assessment of Foreign Currency Risk Premiums," in *Financial Risk: Theory, Evidence and Implications*, St. Louis Federal Reserve Bank, 1986.
[158] Meese, R. and K. Rogoff: "Empirical Exchange Rate Models of the Seventies:

Do They Fit Out of Sample?" *Journal of International Economics*, **14** (1983), 3–24.

[159] Meese, R. and K. Rogoff: "The Out-Of-Sample Failure of Empirical Exchange Rate Models: Sampling Error or Misspecification?" in *Exchange Rates and International Macroeconomics*, ed. by J. A. Frenkel. Chicago: University of Chicago Press for National Bureau of Economic Research, 1983.

[160] Meese, R. and K. Singleton: "A Note on Unit Roots and the Empirical Modeling of Exchange Rates," *Journal of Finance*, **37** (1982), 1029–1035.

[161] Mehra, R. and E. Prescott: "The Equity Premium: A Puzzle," *Journal of Monetary Economics*, **15** (1985), 145–162.

[162] Merton, R.: "An Intertemporal Capital Asset Pricing Model," *Econometrica*, **41** (1973), 867–887.

[163] Miron, J. A.: "Seasonal Fluctuations and the Life Cycle – Permanent Income Model of Consumption," *Journal of Political Economy*, **94** (1986), 1258–1279.

[164] Mishkin, F. S.: "The Real Interest Rate: An Empirical Investigation," in *The Costs and Consequences of Inflation*, ed. by K. Brunner and A. H. Meltzer, Vol. 15 Carnegie-Rochester Conference Series on Public Policy, supplement to the *Journal of Monetary Economics*, 1981, 151–200.

[165] Mussa, M.: "Empirical Regularities in the Behavior of Exchange Rates and Theories of the Foreign Exchange Market," in *Policies for Employment Prices, and Exchange Rates*, ed. by K. Brunner and A. H. Meltzer, Vol. 11 Carnegie-Rochester Conference Series on Public Policy, supplement to the *Journal of Monetary Economics*, 1979, 9–57.

[166] Newey, W. K. and K. D. West: "A Simple, Positive Semidefinite, Heteroskedasticity and Autocorrelation Consistent Covariance Matrix," *Econometrica*, forthcoming (1987).

[167] Nurske, R., *International Currency Experience*. Geneva: League of Nations, 1944.

[168] Obstfeld, M., "Can We Sterilize? Theory of Evidence," *American Economic Review*, **72** (1982), 45–50.

[169] Obstfeld, M.: "How Integrated Are World Capital Markets? Some New Tests," mimeo, University of Pennsylvania, 1986.

[170] Obstfeld, M.: "Peso Problems, Bubbles, and Risk in the Empirical Assessment of Exchange-Rate Behavior," in *Financial Risk: Theory, Evidence and Implications*, St. Louis Federal Reserve Bank, 1986.

[171] Park, K.: "Tests of the Hypothesis of the Existence of Risk Premium in the Forward Exchange Markets," *Journal of International Money and Finance*, **3** (1984), 169–178.

[172] Poole, W.: "Speculative Prices as Random Walks: An Analysis of Ten Time Series of Flexible Exchange Rates," *Southern Journal of Economics*, **33** (1967), 468–478.

[173] Quah, D. and T. Ito: "Estimation and Hypothesis Testing with Restricted Spectral Density Matrices: An Application to Uncovered Interest Parity," National Bureau of Economic Research Technical Paper No. 50, 1985.

[174] Richard, S. and S. Sundaresan: "A Continuous Time Equilibrium Model of Forward Prices and Futures Prices in a Multigood Economy," *Journal of Financial Economics*, **9** (1981), 347–372.

[175] Riehl, H. and R. Rodriguez: *Foreign Exchange Markets*. New York: McGraw-Hill, 1977.

[176] Robichek, A. A. and M. R. Eaker: "Foreign Exchange Hedging and the Capital Asset Pricing Model," *Journal of Finance*, **33** (1978), 1011–1018.

[177] Rogalski, R. J. and J. D. Vinso: "Empirical Properties of Foreign Exchange Rates," *Journal of International Business Studies*, **9** (1978), 69–79.

[178] Rogoff, K.: "On the Effects of Sterilized Intervention: An Analysis of Weekly Data," *Journal of Monetary Economics*, **14** (1984), 133–150.

[179] Roll, R.: "A Critique of Asset Pricing Theory's Tests: Part I: On Past and Potential Testability of the Theory," *Journal of Financial Economics*, **4** (1977), 129–176.

[180] Roll, R.: "Violations of Purchasing Power Parity and Their Implications for Efficient Commodity Markets," in *International Finance and Trade*, vol. 1, ed. by M. Sarnat and G. Szego. Cambridge, MA: Ballinger, 1979.

[181] Roll, R. and B. Solnik: "A Pure Foreign Exchange Asset Pricing Model," *Journal of International Economics*, **7** (1977), 161–180.

[182] Roll, R. and B. Solnik: "On Some Parity Conditions Frequently Encountered in International Economics," *Journal of Macroeconomics*, **1** (1979), 267–283.

[183] Roper, D. E.: "The Role of Expected Value Analysis for Speculative Decisions in the Forward Currency Market. Comment," *Quarterly Journal of Economics*, **89** (1975), 157–169.

[184] Rose, A. K. and J. G. Selody: "Exchange Market Efficiency: A Semi-Strong Test Using Multiple Markets and Daily Data," *Review of Economics and Statistics*, **66** (1984), 669–672.

[185] Rozanov, Y.: *Stationary Random Processes*. San Francisco: Holden-Day, 1967.

[186] Rubinstein, M.: "The Valuation of Uncertain Income Streams and the Pricing of Options," *Bell Journal of Economics*, **7** (1976), 407–425.

[187] Salemi, M.: "Time-varying Risk Premia and the Forward Foreign Exchange Market," mimeo, University of North Carolina, 1986.

[188] Samuelson, P.: "Proof that Properly Anticipated Prices Fluctuate Randomly," *Industrial Management Review*, **6** (1965), 41–50.

[189] Samuelson, P.: "Is Real-World Price a Tale Told by the Idiot of Change?" *Review of Economics and Statistics*, **58** (1976), 120–123.

[190] Sawa, T.: "Information Criteria for Discriminating Among Alternative Regression Models," *Econometrica*, **46** (1978), 1273–1291.

[191] Schmidt, P.: "The Asymptotic Distribution of Forecasts in the Dynamic Simulation of an Econometric Model," *Econometrica*, **42** (1974), 303–309.

[192] Schweppe, F. C.: *Uncertain Dynamic Systems*. Engle Cliffs, NJ: Prentice-Hall, 1973.

[193] Shanken, J.: Multivariate Proxies and Asset Pricing Relations: Living with the Roll Critique," mimeo, University of Rochester, 1986.

[194] Shiller, R. J.: "Do Stock Prices Move Too Much to be Justified by Subsequent Changes in Dividends?" *American Economic Review*, **71** (1981), 421–436.

[195] Siegel, J. J.: "Risk, Information, and Forward Exchange," *Quarterly Journal of Economics*, **86** (1972), 303–309.

[196] Silvey, S.: *Statistical Inference*, London: Chapman and Hall, 1975.

[197] Singleton, K. J. "Risk Averse Speculation in the Forward Foreign Exchange Market: A Comment," in *Exchange Rates and International Macroeconomics*, ed. by J. A. Frenkel. Chicago: University of Chicago Press for National Bureau of Economic Research, 1983.

[198] Solnik, B. H.: *European Capital Markets*, Lexington, MA: D. C. Heath, 1973.

[199] Stambaugh, R.: "On the Exclusion of Assets from Tests of the Two-Parameter Model: A Sensitivity Analysis," *Journal of Financial Economics*, **10** (1982), 237–268.

[200] Stockman, A. C.: "Risk, Information, and Forward Exchange Rates," in *The*

Economics of Exchange Rates, ed. by J. A. Frenkel and H. G. Johnson. Reading, Ma.: Addison-Wesley, 1978.
[201] Stockman, A. C.: "A Theory of Exchange Rate Determination," *Journal of Political Economy,* **88** (1980), 673–698.
[202] Stulz, R.: "A Model of International Asset Pricing," *Journal of Financial Economics,* **9** (1981), 383–406.
[203] Stulz, R.: "Pricing Capital Assets in an International Setting: An Introduction," *Journal of International Business Studies,* Winter (1984), 55–74.
[204] Stulz, R.: "Currency Preferences, Purchasing Power Risks, and the Determination of Exchange Rates in an Optimizing Model," *Journal of Money Credit and Banking,* **16** (1984), 302–316.
[205] Stulz, R.: "An Equilibrium Model of Exchange Rate Determination and Asset Pricing with Non-Traded Goods and Imperfect Information," mimeo, Ohio State University, 1986.
[206] Svensson, L.: "Currency Prices, Terms of Trade, and Interest Rates: A General Equilibrium Asset-Pricing Cash-In-Advance Approach," *Journal of International Economics* **18** (1985), 17–42.
[207] Sweeney, R. J.: "Beating the Foreign Exchange Market," *Journal of Finance,* **41** (1986), 163–82.
[208] Theil, H.: *Principles of Econometrics.* New York: John Wiley and Sons, 1971.
[209] Tryon, R.: "Testing for Rational Expectations in Foreign Exchange Markets," Board of Governors of the Federal Reserve System, International Finance Discussion Paper No. 139, 1979.
[210] Wald, A.: "Tests of Statistical Hypotheses when the Number of Observations is Large," *Transactions of the American Mathematical Society,* **54** (1943), 426–482.
[211] Wasserfallen, W. and H. Kyburz: "The Behavior of Flexible Exchange Rates in the Short-Run: A Systematic Investigation," *Weltwirtschaftliches Archive* 1985.
[212] Wasserfallen, W. and H. Zimmerman: "The Behavior of Intra-Daily Exchange Rates," *Journal of Banking and Finance,* **9** (1985), 55–72.
[213] Westerfield, J. M.: "Empirical Properties of Foreign Exchange Rates Under Fixed and Floating Rate Regimes," *Journal of International Economics,* **7** (1977), 181–200.
[214] White, H.: "A Heteroskedasticity-Consistent Covariance Matrix Estimator and a Direct Test for Heteroskedasticity," *Econometrica,* **48** (1980), 817–838.
[215] White, H.: "Maximum Likelihood Estimation of Misspecified Models," *Econometrica,* **50** (1982), 1–25.
[216] Wihlborg, C. G.: "Currency Risks in International Financial Markets," Princeton Studies in International Finance No. 44, 1978.
[217] Wolff, C. C. P.: "Exchange Rate Models, Parameter Variation and Innovations: A Study on the Forecasting Performance of Empirical Models of Exchange Rate Determination," PhD. Dissertation, University of Chicago, 1985.
[218] Wyplosz, C.: "The Exchange and Interest Rate Term Structure Under Risk Aversion and Rational Expectations," Working Paper, INSEAD, Fontainbleau, France, 1980.
[219] Zellner, A.: "An Efficient Method of Estimating Seemingly Unrelated Regressions and Tests of Aggregation Bias," *Journal of the American Statistical Society,* **57** (1962), 348–368.

INDEX

171

INDEX

FUNDAMENTALS OF PURE AND APPLIED ECONOMICS

SECTIONS AND EDITORS

BALANCE OF PAYMENTS AND INTERNATIONAL FINANCE
W. Branson, Princeton University

DISTRIBUTION
A. Atkinson, London School of Economics

ECONOMIC DEVELOPMENT STUDIES
S. Chakravarty, Delhi School of Economics

ECONOMIC HISTORY
P. David, Stanford University, and M. Lévy-Leboyer, Université Paris X

ECONOMIC SYSTEMS
J.M. Montias, Yale University

ECONOMICS OF HEALTH, EDUCATION, POVERTY AND CRIME
V. Fuchs, Stanford University

ECONOMICS OF THE HOUSEHOLD AND INDIVIDUAL BEHAVIOR
J. Muellbauer, University of Oxford

ECONOMICS OF TECHNOLOGICAL CHANGE
F. M. Scherer, Harvard University

EVOLUTION OF ECONOMIC STRUCTURES, LONG-TERM MODELS, PLANNING POLICY, INTERNATIONAL ECONOMIC STRUCTURES
W. Michalski, O.E.C.D., Paris

EXPERIMENTAL ECONOMICS
C. Plott, California Institute of Technology

GOVERNMENT OWNERSHIP AND REGULATION OF ECONOMIC ACTIVITY
E. Bailey, Carnegie-Mellon University, USA

INTERNATIONAL ECONOMIC ISSUES
B. Balassa, The World Bank

INTERNATIONAL TRADE
M. Kemp, University of New South Wales

LABOR AND ECONOMICS
F. Welch, Texas A&M University, Texas, USA

MACROECONOMIC THEORY
J. Grandmont, CEPREMAP, Paris

FUNDAMENTALS OF PURE AND APPLIED ECONOMICS

PUBLISHED TITLES

Volume 13 (Economics of Technological Change Section)
TECHNOLOGICAL CHANGE AND PRODUCTIVITY GROWTH
by Albert N. Link

Volume 14 (Economic Systems Section)
ECONOMICS OF COOPERATION AND THE LABOR-MANAGED ECONOMY
by John P. Bonin and Louis Putterman

Volume 15 (International Trade Section)
UNCERTAINTY AND THE THEORY OF INTERNATIONAL TRADE
by Earl L. Grinols

Volume 16 (Theory of the Firm and Industrial Organization Section)
THE CORPORATION: GROWTH, DIVERSIFICATION AND MERGERS
by Dennis C. Mueller

Volume 17 (Economics of Technological Change Section)
MARKET STRUCTURE AND TECHNOLOGICAL CHANGE
by William L. Baldwin and John T. Scott

Volume 18 (Social Choice Theory Section)
INTERPROFILE CONDITIONS AND IMPOSSIBILITY
by Peter C. Fishburn

Volume 19 (Macroeconomic Theory Section)
WAGE AND EMPLOYMENT PATTERNS IN LABOR CONTRACTS:
MICROFOUNDATIONS AND MACROECONOMIC IMPLICATIONS
by Russell W. Cooper

Volume 20 (Government Ownership and Regulation of Economic Activity Section)
DESIGNING REGULATORY POLICY WITH LIMITED INFORMATION
by David Besanko and David E. M. Sappington

Volume 21 (Economics of Technological Change Section)
THE ROLE OF DEMAND AND SUPPLY IN THE GENERATION AND DIFFUSION OF TECHNICAL CHANGE
by Colin G. Thirtle and Vernon W. Ruttan

Volume 22 (Regional and Urban Economics Section)
SYSTEMS OF CITIES AND FACILITY LOCATION
by Pierre Hansen, Martine Labbé, Dominique Peeters and Jacques-François Thisse, and J. Vernon Henderson

Volume 23 (International Trade Section)
DISEQUILIBRIUM TRADE THEORIES
by Motoshige Itoh and Takashi Negishi

Volume 24 (Balance of Payments and International Finance Section)
THE EMPIRICAL EVIDENCE ON THE EFFICIENCY OF FORWARD AND FUTURES FOREIGN EXCHANGE MARKETS
by Robert J. Hodrick

Further titles in preparation
ISSN: 0191-1708